Record Keeping

for Organic Growers

Kristine Swaren
and
Rowena Hopkins

A *COG* Practical Skills Handbook

Canadian Organic Growers
Cultivons biologique Canada

Record Keeping
for **organic growers**
First published in 2010

Canadian Organic Growers Inc.
323 Chapel Street · Ottawa
Ontario · K1N 7Z2 · Canada
Tel.: 613-216-0741 Fax: 613-236-0743
www.cog.ca publications@cog.ca

Production
Editor:
Sheila Globus
Design and Layout:
Jean-Michel Komarnicki, JMK Image-ination
Cover design:
Won Pyun, Maria Zaynullina
Front cover photograph:
Firstl*ght Photography, Film & Thought Supplies
Back cover photograph:
Laura Berman, GreenFuse Photos
Photographs:
Page 8 Nat Capitanio
Page 38 and 44 Brian Simpson
Page 67 Les Jardins de la Montagne
Page 71 Rowena Hopkins
Page 81 Alberta Agriculture
Page 82 Penny Marshall
Page 92 Paul Glombick
Project Manager:
Kristine Swaren, Blue Chicory Communications

Library and Archives Canada Cataloguing in Publication

Swaren, Kristine, 1955-
Record keeping for organic growers: COG practical skills handbook / Kristine Swaren and Rowena Hopkins.
ISBN 978-0-9808987-2-9
1. Organic farming--Documentation--Handbooks, manuals, etc.
2. Organic farming--Quality control--Handbooks, manuals, etc.
I. Hopkins, Rowena, 1976-
II. Canadian Organic Growers
III. Title.
S605.5.S92 2010 631.5'84 C2010-907469-6

Printed and bound in Canada

Canadian Organic Growers
Cultivons biologique Canada

CHANGES IN THE ORGANIC SECTOR have been dramatic since COG's inception in 1975. A movement then struggling to be noticed is now a multi-million dollar industry with widespread consumer recognition and national standards backed by federal regulation.

Organic agriculture is now the fastest growing sector in agriculture, and as such it is the most economically and environmentally viable solution for Canada's rural areas.

COG has a significant positive impact on organic growing in Canada through our policy and media work, educational materials, production statistics, scholarships, farmer training, market development, and the grassroots work of our regional chapters.

ORGANIC AGRICULTURE:

- sequesters carbon in the soil and produces food with energy efficient methods
- increases soil organic matter and a diversity of living soil organisms
- improves water quality and quantity
- improves biodiversity
- improves the health of soil, plants, animals, farm workers, consumers
- increases farm financial viability by reducing dependence on inputs and providing farmers a fairer return for their products

COG is a federally registered charity (no. 13014 0494 RR0001). Our members are farmers, gardeners, processors, retailers, researchers and consumers who share a vision of a sustainable bioregionally-based organic food system.

COG's MISSION is to lead local and national communities towards sustainable organic stewardship of land, food and fibre while respecting nature, upholding social justice and protecting natural resources.

JOIN CANADIAN ORGANIC GROWERS

OUR NATURE IS ORGANIC

Contents

Acknowledgments

Few people, especially farmers, enjoy paperwork, and dislike of record keeping is often given as a reason for not becoming certified organic. This is unfortunate because keeping good farm records does not have to be onerous. In fact, farmers who are certified organic say that the paperwork is not much more than what they would do anyway to run the farm. With this *Practical Skills* handbook we hope to build your confidence and knowledge, so that record keeping will actually contribute to your farm's success.

Rowena Hopkins and Roxanne Beavers have offered a record keeping workshop at the ACORN * organic conference for several years. It always attracts a packed room of farmers nervous about the paperwork side of organic growing. We want to acknowledge the inspiration of those who crowded the workshops, as well as those who provided the benefit of their experience.

The farmers interviewed for this handbook all contributed many ideas for successful record keeping. Our thanks go to:

- Lawrie and Margaret Henrey, Klipspringer Farm, BC
- Mark and Sally Bernard, Barnyard Organics, PEI
- Rod and Cheryl Randen, Donalda Springs Organic, AB
- Joscelyne Charbonneau and Sylvain Brunet, Les Jardins de la Montagne, QC
- Tony and Penny Marshall, Highwood Crossing Farm, AB
- Ron and Sheila Hamilton, Sunworks Farm, AB
- Xina Chrapko, En Sante Winery, AB
- Alyson Chisholm, Glen Valley Co-op Farm, BC and Windy Hills Farm, NB
- Lucie Durocher, Ferme Poulet de Grain à l'Ancienne, QC
- Ian Cushon and Jo-Anne Hochbaum, Moosecreek Organic Farm, SK

Several growers, inspectors and certifiers provided constructive criticism of our drafts and helped shape the final book. Thank you especially to Roxanne Beavers and Maureen Bostock.

Finally, our efforts were kept focussed by COG's national director Laura Telford and by our editor Sheila Globus – sincere thanks to both.

Kristine Swaren and Rowena Hopkins
November 2010

* Atlantic Canadian Organic Regional Network (ACORN) www.acornorganic.org

This handbook was funded in part by Agriculture and Agri-Food Canada through the Advancing Canadian Agriculture and Agri-Food (ACAAF) program. We wish to acknowledge the support of the following organizations for making this publication possible: the Investment Agriculture Foundation of British Columbia, the Manitoba Rural Adaptation Council, the New Brunswick Agriculture Council / Conseil Agricole Nouveau-Brunswick, the Territorial Farmers Association and the Yukon Agricultural Association.

Agriculture and Agri-Food Canada (AAFC) is pleased to participate in the publication of this guide. AAFC is committed to working with its industry partners to increase public awareness of the importance of the agriculture and agri-food industry to Canada. Opinions expressed in this handbook are not necessarily those of AAFC.

The authors

Kristine Swaren recently launched a third career operating a certified organic market garden, marketing through home deliveries, restaurants and farmers' markets. With previous experience in writing and project management, she actually enjoys planning and record keeping, but uses her motto "Keep Life Simple" to minimize office time and stress.

Rowena Hopkins has experience both as an organic farmer and organic farm inspector. She is a regular presenter at the ACORN organic conference on subjects such as record keeping and marketing. She was also the Atlantic Canadian trainer for COG's Canadian Organic Standards workshops. During 2009 she took 6 months to WWOOF* across Canada, from BC back to the Maritimes. After that trek she settled in Halifax to manage the Farmers' Markets of Nova Scotia Co-operative.

Kristine Swaren and Rowena Hopkins

* World-Wide Opportunities on Organic Farms

Introduction

How to read this book

If you already do some record keeping for your farm, you're in luck. The good news is that the records you already keep will form the basis of your organic record keeping system. There will be some additional information that you will need to prepare – mostly to provide the "big picture" of your farm – both literally, as in a farm map, and conceptually, as in a description of your overall operation. And that's the extent of the bad news. Really!

This handbook is meant for:

- farmers, ranchers, market gardeners, beekeepers, growers of sprouts and greenhouse crops, and harvesters of wild or cultivated products (e.g. herbs, mushrooms, maple syrup)
- organic, transitional or conventional growers
- producers who do on-farm processing that uses only farm inputs (nothing from off the farm goes into the finished product); large-scale processing is beyond the scope of this book

Record keeping for organic certification is not all that different from record keeping for successful farming. Here's why you'll want to read each chapter:

Chapter 1 — What and why

Records for good farming and records for certification are similar, and both benefit the farm. Chapter 1 introduces the record keeping cycle of organic farming. It also explains the certification context for record keeping: what certification is and isn't, how to choose a certifier, and the steps in the certification process.

Chapter 2 — Farm records

Certification requires records for inputs, activities, and outputs of the farm – whether your operation includes any or all of crops, livestock, or on-farm processing. Many of these are records that you already keep. Here we put them in the context of organic certification and inspection.

Chapter 3 **Applying for certification**

Here's the one you might not have done yet – the organic plan. Typically, your certifier will provide forms for you to describe your operation and your organic methods. In this chapter we discuss the information those forms should contain and the easiest ways to keep that info up to date.

Chapter 4 **Using your records to verify organic integrity**

On inspection day, the organic inspector will observe your farm and examine your records to verify your ability to manage risk and demonstrate traceability—two principles at the heart of organic certification. Here's how to prepare for an easy inspection.

Chapter 5 **Getting better at record keeping**

Whether you are approaching your first inspection or your umpteenth, this chapter provides additional hints and resources for making your record keeping more useful and efficient.

Appendices

A master list and selected templates are contained in the appendices and are also available in electronic format on the COG website (www.cog.ca).

Our aim with this *Practical Skills* handbook is to lessen the anxiety associated with record keeping for organic certification. Whether you choose to certify or not, the record keeping described here will help you transition more easily to organic production methods. Good record keeping also facilitates farm planning and evaluation – which in turn helps to improve your bottom line.

Between chapters, you will find profiles of farms that do record keeping well. There are a variety of operations and sizes, with lots of practical tips based on experience and, yes, mistakes. Whether or not your operation is similar to any of the profiles, they will no doubt provide inspiration.

The chapters contain many examples of specific records and plans; some are from real farms, some are imagined.

By the end of the book, we hope you'll feel a lot more confident in your record keeping abilities. While it looks like a lot of material to cover and a lot of records to keep, we want to help you streamline the process so that the paperwork leaves you lots of time to do what you want to do – farming.

1 What and Why

"Certified organic production can best be described as a comprehensive, rigorous identity-preservation (IP) system. For instance, a field crop operation must keep detailed records on all inputs used, all field activities, harvest, storage and transfer of crops. It must be possible to trace every crop or product from field to fork."

Isaiah Swidersky, organic farmer, ON

Many growers think that becoming certified organic is going to bury them under a mountain of paperwork. In fact, the paperwork you need to be a certified organic producer isn't all that different from what you need to run any profitable operation.

Record keeping provides direct **farm benefits** and shows that you are consistently applying **organic principles.** The record keeping required for organic certification builds on what you already do and on the records you already keep.

Organic certification

Records for certification prove that your practices are organic. They demonstrate the organic integrity of your product – whether plant or animal – from source to sale, through points of potential contamination and commingling.

People want to know where the food they eat comes from. Organic producers have long been ahead of the curve in tracing the origins of our food and ensuring traceability in case of any food safety issues.

By law, it is mandatory to have organic certification if you make an organic claim about your products and if one or more of the following are true:

- you use the Canada Organic logo on your packaging or product
- you sell to international buyers
- you sell across provincial borders
- you sell within a province that regulates organic production

No matter where you sell, the Canadian Consumer Packaging and Labelling Regulations and Food and Drug Regulations can enforce the meaning of "organic" from the Organic Products Regulations. In other words, if you claim to be "organic" in your labelling or advertising, you must be able to prove it, and certification is the only accepted way to do so.

The organic certification system used in Canada and other countries was developed over many years. It is based on international guidelines for certification systems of any kind set by the International Standards

Organization (ISO)[1]. The most important guideline is that verification, decisions, and consultation must be done by separate and independent agencies or individuals.

Your part in this system is to register with a certifier, update your organic plan each year (chapter 3), document your organic practices by keeping records of farm operations (chapter 2), and open your farm for an annual inspection (chapter 4).

In return, you will receive an annual certificate of compliance that you can show to your customers. As well, you may receive support in clarifying the Standards for your operation, and paper forms or computer templates for setting up record keeping. Certifiers review products for compliance with the Permitted Substances Lists, and some may provide brand-name lists of permitted products.

ORGANIC PRINCIPLES

Your records document your organic methods and provide traceability throughout your operation. They show your operation's compliance with the following organic principles, all of which are embedded in the Standards*:

1. Protect the environment, minimize soil degradation and erosion, decrease pollution, optimize biological productivity and promote a sound state of health.
2. Maintain long-term soil fertility by optimizing conditions for biological activity within the soil.
3. Maintain biological diversity within the system.
4. Recycle materials and resources to the greatest extent possible within the enterprise.
5. Provide attentive care that promotes the health and meets the behavioural needs of livestock.
6. Prepare organic products, emphasizing careful processing, and handling methods in order to maintain the organic integrity and vital qualities of the products at all stages of production.
7. Rely on renewable resources in locally organized agricultural systems.

* *Where this handbook refers to "the Standards" it is the Canadian Organic Standards. In June 2009, the Canadian government passed the Organic Products Regulation, making compliance to the Canadian Organic Standards (COS) mandatory and enforceable by the Canadian Food Inspection Agency (CFIA).*

Most international standards are based on the same principles and the record keeping systems required to document compliance are the same. Non-Canadian readers should, however, verify specific requirements with their certifiers or national Standards.

1 Using ISO guidelines, organic inspection procedures and protocols were developed by the International Federation of Organic Agricultural Movements (IFOAM) and the Independent Organic Inspections Association (IOIA). Most national standards are based on this.

Records and plans

Organic certification includes a legal requirement to keep two types of documents: an organic plan and supporting records. These reflect where you are in your operational year; either you are making an annual organic plan as part of your certification application, or you are keeping on-going records about farm activities.

You probably already keep many of the records that you need for certification. Check the chart to see where you comply with the requirements for record keeping, and where you may need to adapt.

Examples of farm and certification records	
Farm records you may already keep	*Additional records for organic certification*
INPUTS	
Receipts for seed, fertilizers, herbicides, pesticides, medicines, livestock	Same receipts; may need list of ingredients to show organic acceptability
Livestock inventory	Same; show source of individual animals and tag numbers (for traceability of organic status)
ACTIVITIES	
Herd health log	Same
Crop spray records	Same
	Farm Journal and Field Activity Log
OUTPUTS	
Sales invoices, Bills of lading, Milk receipts	Same; add affidavits for clean equipment and transport to show no contamination
	Harvest/slaughter yields

As you can see from the chart, most of your current records form the foundation for record keeping for certification purposes. We'll explore this further in chapter 2.

For farmers new to certification, the biggest additional piece of paperwork is an "organic plan.*" It describes your operation and your management practices (chapter 3). You submit your organic plan to your certifier as part of the annual application process. Then, during the season, the organic

** The "organic plan" may have different names with some certifiers: Organic Farm Plan, Organic System Plan, or a similar variation.*

inspector examines your records on-site. In subsequent years, the plan is updated with any new information about your operation.

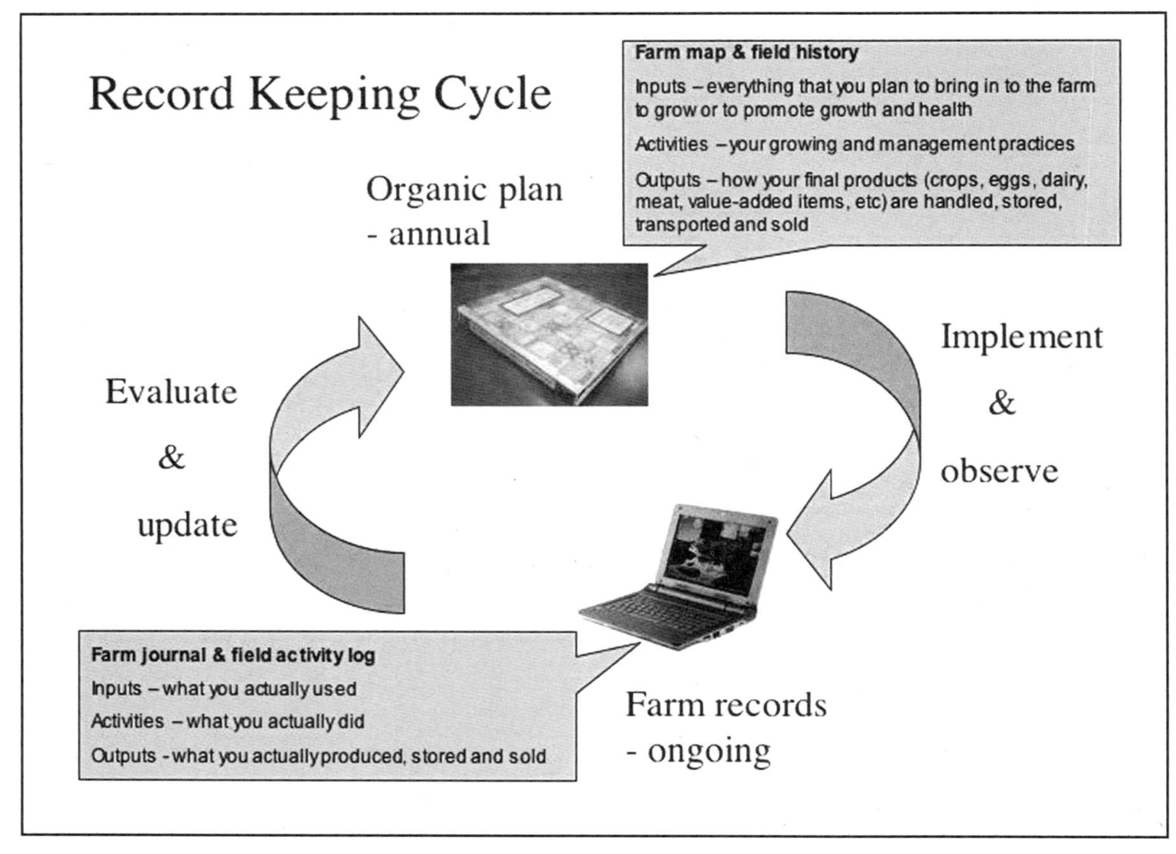

Your organic plan informs the kinds of records that you need to keep and your records feed back into your evolving organic plan.

A record keeping year: compost and manure

What's the difference between plans and records? While the two chapters that follow provide more explanation and examples, this preview demonstrates both the differences and the continuity between plans and records.

Raw or aged manure, leaf mould or compost are common soil amendments applied on organic farms to build soil fertility. You must show your compliance with requirements for organic sources, preparation, application and, possibly, sales – in other words, everything that happens during an operational year.

A record keeping year: the information required

Sources

Show that you have looked first for manure from organic operations, then from unconfined 'naturally raised' livestock and only as a last resort from conventionally raised animals. The Standards do not allow the use of manure from confinement operations where livestock are kept permanently in the dark or cannot turn 360 degrees. If you're using manure from a conventional farm, you will need to document the feed and any vaccines, antibiotics or medications that the animals received. If the animals have been fed GE (genetically-engineered) corn or soy, or if they have received any medications, your organic plan must include an effective composting process for the manure. Certain medications, such as sulfamethazine, do not break down effectively as a result of composting, so you must account for this in your plan. Your certifier must approve the use of any restricted inputs before they are applied.

Preparation

Describe your composting procedure and temperature monitoring. The Standards require any soil amendment described as "compost" from animal sources to have reached the critical temperatures needed to destroy pathogens, because of the potential hazards associated with applying uncomposted or poorly composted manure to organic farm land. If your compost is only from plant sources, you may want to it go through a similar high-temperature process in order to destroy weed seeds, but this is not necessary from a Standards point-of-view. High-temperature composting is also not required for manure that is applied a long time before harvest, more than 120 days for crops in contact with the ground, and more than 90 days for crops with no soil contact.

Application

Show that manure was not applied to frozen or saturated ground; and that sufficient time elapsed between the application of the manure and the date of harvest.

Sale

Show that no contamination of your product happens, by describing your methods and retaining copies of CCME testing for E. coli and heavy metals. Show that your production volumes are reasonable for your inputs and methods. Show that your package labels are accurate regarding organic claims.

The information described in the paragraphs above is recorded either in your plan (your intentions) or records (what actually happened) as follows (page 16):

A record keeping year: where to keep the information

Farm task	Category	Information required	Where to record the information: Organic plan	Records
Purchase compost or manure	Input	■ source ■ ingredients, including the type of animal and the management system it came from (certified organic, feed lot, finishing, confinement, etc) ■ additives or inoculants ■ organic status	■ Inputs source list	■ sources that weren't in the plan ■ receipts ■ labels or MSDS ■ residue analysis or additive specifications
Make compost on the farm	Activity	■ ingredients ■ method(s) that you normally use to make compost (in-vessel, static, aerated, windrow, etc) ■ monitoring for C/N ratio and compost pile temperature ■ residue levels	■ Compost production plan (ingredients, methods, how testing will be done)	■ Compost Log: production record showing ingredients, dates, C/N ratios, temperatures, turning, any changes in methods not in the plan
Apply compost or manure	Activity	■ application dates, rates, fields, crops	■ Field History: under current year, specify fields and crops	■ Field Activity Log or Journal: dates and rates of application to which fields & crops
Sell compost or manure	Output	■ sales and inventory volumes ■ handling, packaging, transport procedures	■ Product handling, storage and transport procedures	■ sales invoices ■ transport bills of lading ■ package labels

Benefits to the farm

Is record keeping worth the effort? Whether for organic certification or not, record keeping can:

- improve productivity and profitability

  Maureen records the date that Colorado Potato Beetles appear each year so that she can begin monitoring on that date the following year. Without early detection, the potato crop would be devastated, decreasing her productivity and profitability.

- make it easier to transfer knowledge to the next generation, farm help, organic inspectors and certifiers

  Jason bought a certified organic farm. Having the records from the previous owner enabled him to start production as certified organic operator immediately – a seamless transition.

- aid financial analysis of the farm and allow for more informed decision-making

  Karen's year-end analysis showed that salad mix was more profitable than spinach. Next year she will grow more salad mix – which can be grown in more parts of her field because it requires less fertility than spinach.

In observing other vegetable growers, one farmer wrote, "When I look around, I see that successful market farmers have a comprehensive approach to record keeping. They have records on harvest, markets, prices received, seed/transplants, fertilizer/soil amendments, crop rotation, sales, expenses, equipment and labour. These farmers – the ones who were able to see the relationship between those detailed records and their total operation – were more profitable.... Good records – kept faithfully over a period of years on your farm – help you plan which early, midseason and late season crops to plant. If you use your harvest records in this way, you should be able to have an adequate quantity and variety of items at your stand each week." [2]

2 Andy King, Pennsylvania, "Record keeping for successful decision making" in The New Farm, Rodale Institute, 2006.

For more on planning for market gardens, see *Crop Planning for Organic Vegetable Growers*, COG, 2010.

Your organic plan contributes to both better farm management and easier certification.

- Farm management benefits
  - helps use available human and natural resources more efficiently
  - anticipates production problems and prepares a range of solutions
  - budgets for the resources, inputs and outputs of the farm, with the goal of becoming more profitable; while not part of the organic plan and record keeping for certification, adding financial data to your records will support cost of production analyses and marketing decisions
- Certification benefits
  - describes how much you can produce with available human and natural resources
  - describes how your management practices comply with organic requirements
  - identifies points where there is a risk of contamination or commingling from non-organic substances, and outlines your risk mitigation strategies

---

HOW CERTIFICATION WORKS

Organic certification is a system based on **Standards**, **procedures** and **records**. The organic sector (producers, processors and sellers) developed the **standards** and **record keeping processes** to ensure the integrity of the entire organic system and to satisfy the requirements of domestic consumers and international trading partners. Each farm develops its own **procedures** (organic plan) that comply with the Standards.

"Process standards operate on the premise that if you start with allowed ingredients and manage them properly, you'll achieve a desirable outcome. This approach is a lot more flexible than relying upon a product standard in which some sort of testing is used to verify that the system is working. Think about this distinction in regards to raising a child: isn't it easier to teach the right values than to spend your time looking over his or her shoulder?"[3]

3 Mark Keating "*Transitions: Thoughts on Organic Food & Farming*" Acres USA, November 2008.

The organization that independently verifies that you are following the Standards is called a Certification or Certifying Body (abbreviated CB), or Certifying Agency – and also just Certifier, which is the term used in this book. All certifiers are overseen by Conformity Verification Bodies. These bodies are audited annually by the government to assess their ability to do their job.

You must register with a certifier and use their services to verify your compliance with the Standards. Certifiers are not government agencies, and neither are the people who write the standards. This allows for the independence of those who set, verify and enforce the standards.

The Standards specify the information that you need to keep in your records; COG's Guidance Document further explains the legal requirements for the Canadian Standards (see the excerpt in Appendix D). The instructions and forms that are included in this book are generic. Most certifiers have their own templates, so it is best to register with a certifier and determine their particular paperwork requirements.

Choosing a certifier is one of the most important business decisions that you can make as an organic producer, so it is worth spending the time and energy to do your research. Ask nearby farmers for advice, find out which certifiers operate in your region and market[4], and consider working with a certification consultant[5]. The following steps should help.

STEP 1: KNOW YOUR PRODUCT AND MARKETS

You will want to contract with a certifier with accreditation in the markets that you wish to reach:

- if you are growing and selling within your own province, provincial regulations apply
- if you are selling to buyers in another province, Canada's national Organic Product Regulations apply
- if you have international buyers, you must meet the certification requirements of the importing country

---

4 For certifiers operating in Canada, check the list on the CFIA website (see Resources).

5 An organic transition specialist or consultant provides guidance regarding your preparedness for certification and an appropriate choice of certifier. Contact your provincial or regional organic organization and ask to be referred to an appropriate person. These are not usually the same as government organic extension specialists (who specialize in farming, not certification) and some charge a fee while others hold funded positions with not-for-profit organizations.

Equivalency agreements negotiated between countries enable producers to certify only to the Canadian Organic Standards (COS). The Canada Organic Office (COO), part of the Canada Food Inspection Agency (CFIA), is working to increase the number of countries that recognize the Canadian Organic Standard. This international harmonization of standards will make it easier to trade globally. Where equivalency agreements are not yet in place, many of the certifiers that operate in Canada have accreditation in those countries.

STEP 2: KNOW WHICH REQUIREMENTS APPLY

You will need to obtain – and read – a copy of the Canadian Organic Standards[6] (COS) to understand the requirements that apply to your operation. The Standards come in two parts:

- **CAN/CGSB-32.310 ORGANIC PRODUCTION SYSTEMS GENERAL PRINCIPLES AND MANAGEMENT STANDARDS**, and
- **CAN/CGSB-32.311 ORGANIC PRODUCTION SYSTEMS PERMITTED SUBSTANCES LIST**

CAN/CGSB-32.310 includes the standards for production methods for:

- crops
- livestock
- apiculture
- maple products
- mushrooms
- sprouts
- greenhouse crops
- wild crops

CAN/CGSB-32.311 identifies the substances allowed for use in an organic operation. If a substance is not on the list, the odds are that it is not allowed. Permitted substances are organized into tables for crop and livestock production, processing and cleaning.

Read these two documents thoroughly to determine which sections apply to your operation. Identify potential factors that might prevent or delay your certification. Discuss these with your certifier before making any changes to your practices.

---

6 Available from the Canadian General Standards Board by calling 800-665-CGSB or online at (English) http://www.tpsgc-pwgsc.gc.ca/cgsb/on_the_net/organic/index-e.html

PARALLEL AND SPLIT PRODUCTION

PARALLEL PRODUCTION

If you are planning a phased transition to organic, note that "parallel production" is not allowed under the Standards. Parallel production means that your organic and non-organic products are indistinguishable; for example, two fields of Eston lentils, one organic and one conventional.

SPLIT PRODUCTION

With very careful record keeping, you can maintain a "split operation" where some production is in transition and some is non-organic, as long as the end-products are distinct; for example, red and green varieties of lentils, or lentils and potatoes.

RECORDS

Your records must reflect the extra care you take to keep the two growing systems separate and the organic crop or livestock uncontaminated. The Standards require that the operator develop a plan to eventually transition all elements of the farm to organic.

STEP 3: CHOOSE A CERTIFIER

Use some or all of the following questions[7] to evaluate the certifier's suitability to your own business and philosophical needs:

- What experience do they have with your particular crop, market, livestock type, or farm size?
- Do they certify to the international standards that will allow you to export to your country of choice?
- When is someone available by phone to answer questions about certification or the Standards? What is the turnaround time on returning calls?
- Do they provide record keeping forms and templates? Paper or electronic?
- What kind of support do they offer? For instance, some certifiers have a club structure that offers peer support and educational opportunities.
- Does the certifier have its own brand name directory (useful for determining whether your inputs and products comply with the PSL)?
- What is the fee structure? Annual fees can be a flat rate or based on the size of your operation or gross income; ask also if there are additional fees for inspection or for dealing with noncompliance issues.

These questions, together with questions of your own, should give you a good idea of whether the certifier:

- can provide direction and clarification on regulatory questions and on the Standards

7 Adapted from Organic Producer online magazine "Ask the certifiers" Q&A www.organicproducer.com

- has a fee structure that is realistic, sustainable, and fair
- is knowledgeable about crops, conditions, and markets
- has a philosophy and mission consistent with helping producers be organic and sustainable

You should also feel good about the relationship you have with your certifier; how well do you communicate with each other? Talk to other farmers to find out what their experiences have been and get their recommendations.

WHAT TO EXPECT FROM A CERTIFIER

The initial application is the most work; subsequent years require only updates to the original. Here's what you need to know about working with your certifier.

- The certifier will provide you with application forms and a selection of supporting documentation templates. These are designed to help you demonstrate organic integrity and traceability.
- Your certifier may offer newsletters, print and/or web resources, and licensed use of their logo on your labels. They are also authorized by the Canadian Government to control the use of the Canada Organic logo.
- If your farm is found non-compliant, your certifier may help clarify the Standards as they apply to your operation, so that you can resolve the non-compliance problem.
- The certifier's employees and inspectors are professional and impartial; they operate under strict confidentiality and conflict of interest guidelines. Organic inspectors are often independent contractors and may work for more than one certifier.
- You'll get the best service from your certifier if you provide enough information for the inspector and certification committee to easily assess your operation. Demonstrating organic integrity is the goal shared by farmer and certifier.

STEPS IN THE CERTIFICATION PROCESS

(Forms or documents are indicated in bold.)

Step 1: Applicant initiates the organic certification process with a certifier, requesting an application

Step 2: Certifier screens applicant to determine eligibility and which organic plan Application and attachments are appropriate to the operation.

Step 3 Applicant completes **organic plan, Application and attachments.**

The package is returned to the certifier along with the **Certifier Licence Agreement** (or contract).

Step 4: Certifier screens all application documents to determine applicant's eligibility (ability to comply with the standards) and to assess clarity and completeness of the application.

Step 5: Certifier assigns an organic inspector, who conducts an on-farm inspection and submits a report to certifier.

Step 6: Certification Committee reviews complete organic plan (Application, Farm Plan and attachments, Organic Inspection Report), and makes decision on organic certification status.

Step 7: Certifier issues a contract stating organic certification status. A notice to comply is sent at this time if all organic conditions have not been met. Organic status will not be issued until all major non-compliant issues have been corrected.

Step 8: Applicant signs and returns **Organic Certification Contract** and implements all conditions. (Step 3 is the signed contract with some certifiers.)

Step 9: Upon receipt of the Organic Certification Contract and when all conditions have been met, applicant receives their **Organic Certificate.**

Step 10: Annual Re-Certification – Applicant submits **Re-Certification Application** and appropriate attachments. Repeat steps 3-9.

Source: Adapted from IOIA Inspection Manual

"Canadian Organic Growers (COG) encourages all farmers, regardless of where they choose to market their products, to consider becoming certified. Certification will provide consumers and our trading partners with the integrity and legal authority they seek. It also puts all organic farmers regardless of the size of their operation on the same team in solidarity. By choosing to certify, organic farmers support one another.

COG will continue to work with governments to ensure that the process is designed with farmers in mind and that the certification process is accessible, easy to use and affordable.

US experience has shown that organic labels offer significant marketing advantages even for farmers who market locally. There is now widespread brand recognition of the USDA organic label, and market research shows that consumers preferentially purchase products carrying the label."

—Laura Telford, national director, Canadian Organic Growers

KLIPSPRINGER FARM

LAWRIE AND MARGARET HENREY, GIBSONS, SUNSHINE COAST, BC

FARM:
Market garden, 2.5 acres (1 ha) in production

ORGANIC CERTIFICATION:
PACS

FIRST CERTIFIED:
2003

RECORD KEEPING SYSTEM: Paper

KLIPSPRINGER FARM is named for a South African deer that stands only 58 cm (22 in) tall but can jump as high as 7.6 metres (25 ft). On Lawrie and Margaret Henrey's farm however, the 2.1 meter-high (7 foot) fence is more than enough to keep the local variety of deer out of their row crops, berries and fruit trees.

The Henrey's main crops are lettuce, berries, garlic, peaches, cherries, tomatoes, fennel, peas, beans (favas and green), and herbs. The farm supplies two local organic food stores, a local supermarket's organic produce section, and a local restaurant. They also prepare custom-picked orders for regular customers and farmers' markets. Lawrie and Margaret have always grown organically, and became certified in 2003. Here's how the Henrey's describe their operation:

"All our beds are raised to ensure good drainage, ease of working, and soil warming. We have a warm-germination greenhouse with automatic air temperature control and thermostatically-controlled heating mats. To germinate seedlings we use various sizes of planting blocks. There is an adjoining cool greenhouse for hardening off seedlings before transplanting, and a small greenhouse for winter growing."

"We maintain a hand-written plan of all the growing beds and areas, which are numbered on the plan and on permanent markers on each bed. We made a template that we have found very useful for tracking crop rotations. Each numbered bed has its own page that shows 7 to 10 years of crop rotation. (The example for Bed 9 shows crops from 1999 to 2008.) The recorded information shows crop name, botanical name and family name in adjoining columns, so it's easy to see whether you're making a good rotational choice when planning for the following season."

"We plant cover crops as soon as the main crop is harvested, and after we've weeded and composted. We use greensand, alfalfa meal/pellets, kelp meal, lime or crab meal, and rock phosphate as amendments. Having all this information on one form makes it easy to show the Verification Officer at the annual inspection and to complete the records required by the certifier."

Bed # 9	Size 195" X 52"		Bed #	Size		Bed #	Size	
YEAR 1999			YEAR 2004			YEAR 2007		
CROP GARLIC	Botanical name	Family name	CROP	Botanical name	Family name	CROP	Botanical name	Family name
FALL 1999	ALLIUM	LILIACENE	STRAWBEARIES			SEEDLINGS		
GARLIC	SATIVUM		Good CROP			FROM GREEN-		
FALL 2000			Too MATTED			HOUSE –		
COVER MIX			2005			LETTUCE	LACTUCA	COMPOSITAE
FALL RYE			PLANTS			SALAD MN	sativa	
BUCKWHEAT			THINNED			SHADE CLOTH		
WHITE/RED			MAR 2005			SEPT. 07		
CLOVER			APL -2005			INFRA-RED		
MAY 2001			RUNNERS			SOIL COVER		
ARTICHOKES	CYNARA		TRANSPLATED			2008		
FALL 2001	SCOLYMUS		TO U/S BEDS			PLAN CUKES	CUCUMB	CUCUR-
HAIRY VETCH			AUGUST 2005				SATIVUS	BITACEAE
FALL RYE MIX			S-BERRIES OUT					
MAY 2002			SEPT. 2005					
TRANSPLANTS			COMPOST +					
BROCCOLI	BRASSICA OLERACEA	CRUCIFERAE	ALFALFA					
LETTUCE	LACTUCA SATIVA	COMPOSITAE	MEAL					
FAVAS	PHASEOLUS	LEGUMINOSAE	OCT. 2005	ALLIUM	LILIACENE			
SEPT 2002			GARLIC	Sativum				
COMPOST &			JULY 2006					
FALL RYE			GARLIC					
2003			HARVEST					
STRAWBERRY	FRAGARIA SPP	ROSACEAE	AUG 2006					
RUNNERS			COMPOST/OATS					
PLANTED			HAIRY VETCH					

"For our crops, we have a weather-proof binder with a write-up of the planting requirements for each crop, filed alphabetically by crop name. It accompanies us to the greenhouse, germination house and field. Each crop entry (see sample on Beets) includes plant and row spacing, pH, amendments, companion and under-sowing preferences, pests that affect that crop and pest control methods (we prefer to use row covers and plant herbs and flowers to attract beneficial insects and birds). The write-up on the individual crops is continually updated as new information comes to hand. With two employees, the crop binder is also useful for communicating the timing of actions for pest control, soil enrichment, and other plant care."

"We use another binder to record each year's seed orders and supplier acknowledgements. Our sales register is summarized from copies of numbered invoices given to all regular customers who call at the farm, and from one daily entry for casual customers (unless invoiced) and one record of sales after each farmers' market."

"For inspection, we provide the results of our annual compost test by the Soil Food Web. We also get an *E.coli* test every second year which costs $200 and always shows no *E.coli* present. Nonetheless, we do it for the satisfaction of confirming that no *E.coli* is present in our aerobically produced compost."

Beets

Vulgaris subsp esculenta. Family: *chenopodiaceae* (Spinach family)

Soil pH 6.8 – 7 **ideal. Adding lime usually beneficial on Sunshine coast.**
Eliot Coleman sows 4 fruits inside in each 2 ½" **multi-plant soil block (see The New Organic Grower)** for early crop and transplants the blocks, at 4 weeks outside, 6" **apart in** 16" **rows.** It is very important not to allow the blocks to dry out. Thinning is not usually necessary as the dominant seed in each fruit prevails. Seeds are actually **fruits containing up to** 4 **seeds** so thinning is important, unless the soil block system is used. If not, the rows should be 18" apart but early planting and baby beet planting can be in rows 12" apart with seedling at 2" spacing apart

Sowing: avoid cold soil; wait until end of April and sow seeds at 2 week intervals outside in sandy soil from last week April until mid July, when soil temperature is in the germination range of 10 – 26 degrees C (50 – 80F), or sow early spring in cold frames about 5-6 weeks before transplanting. They will tolerate some shade.

Transplant in 12" rows 4-8" apart for larger beets, 2" apart for greens, 3" apart for baby beets. Mulch between rows.

Boron. Black cankers in roots usually indicates lack of **boron**. It may need supplementing. Use pelleted borax with 10% elemental boron content, or dissolve 1 tablespoon (**<u>no</u> more**) of **borax** to 4 litres of water and spread evenly over 100 square feet of soil

Sow/transplant in **deeply dug soil (at least** 8"**) with compost above and ½ cup organic fertilizer** mixed into each 5' of the planting row soil **below the seed**. Beets are moderate feeders. Avoid fresh manure which makes the root grow side shoots instead of one round beet. Add compost to soil as a top dressing when beets half grown.

Pale rings are due to **dry soil**; regular water is important.

Sow seed ½" deep, 1" apart and thin (and eat) carefully to 5" apart when seedlings are 2" tall.

Cover beds with **floating row cover** to prevent **flea beetles** (their eggs hatch into wire worms)and the small yellow **leaf miner fly** from laying eggs and prevent their maggots from eating leaves.

Harvest at any convenient size and cool quickly with leaves. If leaves are not cut you can harvest uniform round beets. Up to a third of the leaves off each plant can be cut and used as cooked greens.

Certified Organic Red Ace seed is available from Johnny's (who say it is **the best beet**) and WCS. It has fast spring growth; deep red roots are uniform, sweet and tender. Eliot likes best Red Ace and Vermillion. **Bulls Blood** leaves are the most attractive in salads, and darken as leaf grows larger. **Early Wonder Tall Top** is good at all seasons and a very good early variety with quick growth in chilly soils. The tall tasty green leaves with red stems age good eating.

Rotation. Beets usually follow **beans or cabbage or kale** and are followed by **greens.**

2 Farm records

Every tractor cab has a notepad and all members of the team have had to develop the discipline for record keeping... We all recognize that proper record keeping is not a restraint; it's a way to improve our practices.

Guy Gautier, organic farmer, QC

Your records describe what actually happens on your farm and deepen your understanding of the interactions of the soil, plants, animals, pests, diseases and growing conditions that are specific to your operation. The information they contain will help you be proactive rather than reactive.

Your records show how your operation complies with the Standards under normal circumstances and under circumstances caused by things such as weather, pests, diseases, or contamination from neighbours.

In this chapter we will discuss, in detail, a selection of records that certified organic farmers need to keep. We'll also provide examples and offer tips for making record keeping simpler so that you can achieve the following:

- demonstrate that you followed organic principles compliant with the Standards
- document specific data such as quantities, locations and dates
- provide traceability, backwards or forwards, from source to sale including tracing all activities between sales of farm products and the sourcing of the seeds or livestock that generated those products
- provide information that will help you improve your organic plan in future
- enable certifiers to carry out the certification process more efficiently thus saving you time and money

Virtually every kind of record that is relevant to your farm operation is also relevant to organic certification and vice versa. Some examples:

Type of record	For organic certification	For organic farm management
Harvest records	Establish whether yields are within the expected range for the area planted (demonstrating that there has been no buying and re-selling of non-organic product.)	Monitor yields over time to establish which the most productive and most valuable crops are.

Herd Health Records	Ensure that the use of veterinary drugs conforms to the Canadian Organic Standards and provides an overview of herd or flock health for the organic inspector.	Determine whether problems lie with individual animals or with overall herd health.
Production line clean-up records	Demonstrate that organic integrity and health and safety standards are being met.	Create a standard process with instructions and a checklist that any staff member can follow without supervision.

You don't need separate files for organic certification as long as all records relevant to certification are readily accessible at the time of inspection.

Keep it simple and you'll be more likely to keep good records.

What are records?

Records: Any information in written, visual or electronic form that documents the activities undertaken by a producer or a person engaged in the preparation of organic products, in accordance with this standard.... – COG Guidance Document

Your records describe what happens on your farm, with specific information such as dates, quantities, and locations. Think of them in terms of inputs, activities and outputs.

Records typically consist of lists, charts and tables plus an overall calendar or journal for recording events and observations. You will also keep "supporting" documentation such as labels, receipts and affidavits.

Depending on your operation, here are some examples of the kinds of records you might keep.

Type of record	Crop operation	Livestock operation	On-farm processing operation
Inputs – what you actually used	Suppliers list	Purchased dairy goats register	Ingredient list
Activities – what you actually did	Composting record	Livestock health record	Processing log

Outputs – what you actually produced, stored and sold	Harvest records	Egg counts & sales records	Inventory records
Supporting documents	Seed labels	Letter re: source of livestock	Statement regarding inert ingredients

The Farm Journal is the basic operational record on most farms, but summary forms, such as the examples above, can be helpful for finding information quickly. Certifiers often provide templates for records that will help with certification and with farm management.

The Records Master Chart in Appendix A lists common forms for record keeping, with templates provided in Appendix B. Use these samples or your certifier's forms as a basis for your own records.

FARM JOURNAL

"It is important to develop a daily discipline with record keeping. Get used to the fact that this aspect is essential and is part of normal work, not extra work."
Simon Halde, organic farmer, QC

The farm journal (or calendar) is where all activities are recorded. Make your entries daily if possible. You'll have everything in one place and can use the information later to update specific documents for your inspection or for next year's application.

Get yourself a small notebook in pocket or glove box size, whichever is more convenient. Label each page with a field number. Keep this handy at all times.

At coffee break, or as it happens, write down what you saw or did on the fields. Include everything: when and what was seeded (if the seed was purchased, keep receipts and labels), weed or pest problems and what you did about them, soil improvements, and when the crop was harvested, your yields, and the number of the bin that you stored the crop in.

Remember to date everything. Keep receipts and labels for all inputs. If you are doing something special with part of a field, such as setting aside a portion for wildlife, mark it down in your book. This will be invaluable when you go back to filling out your field history form and also when the organic inspector starts asking questions.

Do this for both your conventional and your organic crops. The inspector will want to be sure that you are keeping good records and that there is no chance your conventional crops will get mixed in with your organic crops. If you are using the same piece of equipment for both conventional and organic crops (e.g. truck, combine, auger) then write down in your book what you did to clean each piece of equipment before putting organic crop in it. The same goes for any custom work you may have done. Make sure that only clean equipment touches your crop and write down what you did to be sure[8].

Some growers keep a paper journal. Others find a computer more practical. (See chapter 5 for resources for computerized record keeping.) There is no right or wrong way to keep a Farm Journal as long as you can update it regularly and can extract the information you need when you need it.

Consistency is crucial in terms of where you record information and how often you do it.

ELEVEN TIPS FOR YOUR FARM JOURNAL

The main challenge is just taking the time to sit and record what was done – it is important to plan this into your day! You have to write it down today or it won't get written.
– Lorne Jamieson, organic farmer, ON

1. Update your journal regularly so that you don't forget important details. Treat it as a time of reflection at the end of the day.
2. Keep notes on a notepad or clipboard as you work and transfer them to your journal at the end of the day or week. Just don't lose that notepad!
3. Buy notepads in bulk in multicoloured packs, and assign each staff person their own colour for keeping field notes. At the end of the day, empty notes from pockets and file them or transfer the information to the Farm Journal.
4. Keep your journal in one location so that you don't have to hunt for it.
5. Have a spare blank journal ready for when you have to start a new one so you won't have to write on envelopes or scraps of paper.
6. Use a pen or highlighter to colour code key information such as crop, livestock or field names. It'll be easier to find the information later.
7. Be consistent in how you sequence information. For instance, start each entry with the field name and crop name.
8. Use sticky-notes to flag pages that you refer to frequently – these are usually lists or maps. Or, keep them together in one "reference" section of the journal.
9. Start a new journal each year and write the year prominently on the cover and down the spine. Store your old journals in one location.
10. Remember to back up any computerized information **regularly** onto a memory stick, disc, external hard-drive or online storage (such as dropbox.com or Google Docs) so that if your computer dies, your records won't die with it. Add "backup files" to your weekly or monthly to-do list, and keep a copy offsite in case of a total disaster in the farm office.

8 *adapted from OCIA "Six Steps to an Easier Inspection"*

11. Keep your records in a format that you enjoy using and you will be more likely to keep them current.

Inputs:
SOIL AND AMENDMENTS, TREATMENTS, SEED, LIVESTOCK, PROCESSING INGREDIENTS, CLEANERS

Input records cover anything that you bring on to your organic farm–from seed to feed and lime to lime juice. For purchased inputs, you already keep receipts for tax purposes. For inputs that are produced on-farm, such as saved seed, straw or compost or livestock that you breed yourself, you *must* keep production records (examples will be in the "Activities" section below).

Whenever you purchase a product, you must keep sufficient documentation from the following list to demonstrate that it is organic or an acceptable input:

- seed tag or product label
- invoice showing organic status of supplier
- copy of the supplier's organic certification
- non-GE (genetically-engineered) statements printed from websites or cut out of brochures and catalogues (this could be for seed or products such as biodegradable plastic mulch or packaging)
- ingredient lists cut from the sides of packets, especially new or unusual products that the organic inspector might not recognize
- Material Safety Data Sheet (MSDS) from the manufacturer
- brochure, label or manufacturer's letter to show that packaging materials such as crates, cardboard boxes, berry pots and bags are food grade, fungicide- and GE-free

A few examples of Input records follow. You can use the master chart in Appendix A as a checklist when setting up your records for your type of operation.

CROPS EXAMPLE: PURCHASED SEEDS RECORDS

Jeff plans to seed field #303 to soybeans. He wants S03-W4 but cannot find organic seed at the local feed store. Jeff will be required to submit an Organic Seed Exemption Application prior to sourcing the seed conventionally. (The Canadian Organic Standards require that a search be conducted checking with known suppliers of organic seed prior to using conventional seeds.) Approval for the exemption must be obtained prior to seeding date. As support for his Organic Seed Exemption application, Jeff will log his efforts to obtain organic sources by saving letters and records of phone calls, email, or Internet searches.

Julia operates a market garden and grows a wide selection of crops. Each year she creates a seed search document that lists her seed suppliers, the cultivars that she purchases and whether or not the supplier carries organic seed for each cultivar she wants.

SAMPLES OF SEED SEARCH RECORDS

Certified Organic Oat Seed Search

COMPANY	WHEN CONTACTED	RESPONSE
Farmers Co-op, Leduc	April 2	Looking into it. None in stock.
Murray Organic Farm	April 2	Only has enough for his own needs
Homestead Organics	April 3	Certified Organic oat seed available, but shipping is prohibitive. Looking into combining a load with neighbours

Market Garden Seed Search

CULTIVARS	SUPPLIERS					NOTES
	Veseys Seed	William Dam	Johnnys SS	High Mowing Seeds	West Coast Seed	
Butternut Squash	x	o	o	x	o, yes	
Napoli Carrots	x	o	o	x	o, yes	
Provider Beans	o	o	o	o, yes	o	
Premium Crop Broccoli	–	–	–	–	x, yes	This cultivar performs best under local conditions; trialing two organic cultivars from High Mowing this year as potential alternatives.

– = not available at all

X = not available certified organic

O = available organically

yes = supplier purchased from

LIVESTOCK EXAMPLE: LIVESTOCK REGISTER

The livestock register demonstrates which of the animals were born on the farm after certification commenced, which dairy cows were transitioned as a herd, how long each of the breeding animals has been under organic management, and which animals cannot be slaughtered for organic meat.

For each new livestock purchase, Janet has a receipt complete with date, organic status, Livestock Identification Number and certifier. There is one adult meat animal that lacks documentation, so it is designated non-organic and therefore only useable as breeding stock.

Every two years she exchanges some of her herd with other organic breeders in order to maintain genetic diversity. She provides documentation that proves the organic status of her animals, and receives similar proof about the incoming animals. This can take the form of annotated receipts or a letter detailing the exchange.

Date arrived on farm	Livestock identification number.	Organic status	Documentation	Purchased from
26/08/09	257 398 972 257 398 973 257 398 974 257 398 975 257 398 976 257 398 977	Certified organic by BCARA	Receipt of purchase. Transaction Record.	Millcreek Organic Farm, Abbotsford, BC
30/10/09	257 398 136 257 398 137 257 398 138 257 398 139 257 398 140	Not certified, but 'Naturally' raised with no antibiotics.	Receipt of purchase. Affidavit.	Rising Sun Natural Dairy, Sherbrooke, QC

PROCESSING EXAMPLE: SALSA INGREDIENT LIST

Gabriella's salsa is made on her certified organic farm with most of her own produce, but a few ingredients are purchased. For each batch made, she keeps a log that shows the sourcing of all ingredients and their organic status. She keeps receipts and labels for purchased ingredients to prove their organic status and certifier.

Ingredient	Supplier	Certified organic by	Lot number	Amount in recipe	%
Roma Tomatoes	On farm	Ecocert	257	200 lb	67.00%
Jalapeno peppers	On farm	Ecocert	260	2 lb	<1%
Corn	On Farm	Ecocert	250	30 lb	10.00%
Green Peppers	On Farm	Ecocert	262	50 lb	17.00%
Red wine vinegar	Co-op	Bios Sri	0198AD	13 lb	4.50%
Brown Sugar	Co-op	USDA	GHY23	5 lb	1.50%
Salt (non-iodised)	Co-op	n/a	-	0.25 lb	<1%

7 TIPS FOR INPUT RECORDS

1. Remove package labels and keep them on file; you don't have to keep entire tote bags or other large containers.
2. Remove labels before they become sun-bleached or covered in mud.
3. Stick small seed packets into a scrap book or in large envelopes, labelled by year and, if necessary, by crop type.
4. Keep receipts and delivery slips in an 'inputs' file; these often contain detailed product information.
5. Keep a sample label from each batch of vaccines (same lot number) rather than from every bottle.
6. Store all large unfinished packages in the same place so that their labels can be checked easily by the organic inspector.
7. Keep an affidavit from the neighbour who brings over a load of old manure, describing its source, organic status, age and content.

Activities:

FIELD WORK, HEALTH MANAGEMENT, ON-FARM PROCESSING

"Activities" include what you did, when and where you did it, and what you used. Examples include field work such as harrowing, tilling and plowing; applying inputs such as soil amendments, foliar sprays, compost teas or organic pesticides. Activity records often correspond to repeated processes that are described in your organic plan. For example:

Annual organic plan	Daily activity records
Compost recipe, method and sources of inputs	Composting Log including dates when material was added, dates when the pile was turned and what temperatures were reached.
Description of herd health management including products and methods of application	Herd health log including dates of treatments and animals treated
Product Recipe and Process Description	Processing and Packaging Log

CROPS EXAMPLE: FIELD ACTIVITY LOG

Nick keeps a chronological log for each field, giving an at-a-glance summary of all crops, inputs, field work and observations.

FARM NAME NORTHWINDS NINETY LAND LOCATION SE 12 - 63 24 W4
FIELD NUMBER # 2

STATUS: O - Organic (3 yrs. Chemical free), T - Transition (2 or 1 yr chemical free), C - Conventional

DATE	O/T/C	ACRES	SOIL	CROP	PREHARVEST		HARVEST		BIN # or	LOT #	COMMENTS
			INPUT		CULTIVATION	SEEDING	CUTTING	COMBINE	Storage		
1995	O	22	Fallow								
95/09/04	O	22		Rye	deep till	Sowed Fall rye					using as weed control
96/09/12	O	22		Rye				900 b.	3	9623R	
97/05/12	O	22		oats			Swathed then Combine	450 b 100 b	2 4	97223	
98/04/22	O	22		/	deep Till + harrow	Rye	mowed twice.				Sowed rye as green manure
99/04/27	O	22			Plowed						

A more extensive format for a field activity log might also include soil tests, and notes about weed, pest, and disease problems.

CROPS EXAMPLE: COMPOST RECORDS

Don composts all the dairy and poultry manure on the farm; the straw is his own as well. His log might look like this:

Composting Log

Date	Ingredients added	Pile turned?	Temp (°C)	Notes
10/06/09	Manure from Hen Barn with straw bedding	No	67	40 buckets. Overheating. Need to add carbon source
12/06/09	Wood chips from lumber yard	Yes	53	Half a small dump truck (5 cubic yards)
16/06/09	Dairy manure from winter yard with some straw	Yes	58	8 cubic yards
18/06/09	-	Yes	57	
20/06/09	-	Yes	59	

LIVESTOCK EXAMPLE: MEDICAL TREATMENT LOG

Laurie raises sheep on pasture. If the field rotations are not managed carefully or the pasture is wetter than usual, there may be trouble periodically with the sheep developing high levels of internal parasites. Her medical treatment log gives the name of the product used, the date of treatment, the animal(s) treated, and the method and reason for treatment. The organic inspector will check the log to ensure that no slaughter animals have received more than the allowable number of treatments.

Date	Product used	Livestock tag	Treatment method and amounts	Reason for treatment
6 Mar 2008	Diatomaceous Earth	All lambs	Added to creep feed	Promote healthy gut to resist parasite infections.
2 Sep 2008	Ivomec	B52	Single treatment	High numbers of hairworm eggs. Ivomec recommended by vet (in writing).

PROCESSING EXAMPLE: FRUIT WINE

En Sante Winery in Alberta documents what happens to each batch of wine they make. This not only helps them produce a consistently high quality product, it also allows them to demonstrate traceability. The Batch Log below summarizes what happens to each batch from the time it is put down to ferment to the time it is bottled. The batch number can be cross-referenced by the organic inspector as well as by the winery as required.

Each of the wines has a standardized recipe. These are kept in a separate binder and are simply replicated each time a batch is made. As an activity record, the Batch Log contains information about how much was produced, the date of the processing, the lot number, who processed that particular lot, and any variations from the normal recipe or procedure.

Fruit Wine Batch Log

	Batch Number: 07-*SA*-106
Wine Name: *Saucy Saskatoon*	Starting Date: 2007 *Aug* 13
Quantity of Base: *saskatoons – 200 kg* *See recipe for other details*	Water: 600 *litres* Yeast Type and Amount: *Lav* 225 *gr*
Initial Specific Gravity: 1.078	Fermentation Temperature: 19 – 26 *C*
Date moved under air lock: 2007 *Aug* 18 Specific Gravity: 1.032	
1st Racking Date: 2007 *Aug* 26 Comments: *S.G* 1.018	2nd Racking Date: 2007 *Sep* 17 Comments: *S.G.* 0.095
3rd Racking Date: 2007 *Oct* 11 Comments:	4th Racking Date: 2007 *Nov* 29 Comments:
Filtration Details: *BECO SS*	Bottling Details: 2008 *Apr* 10 824 bottles

FOUR TIPS FOR ACTIVITY RECORDS

1. Keep your forms on a clipboard hanging on a nail by the barn or greenhouse door or in a folder on a shelf in the processing facility.
2. Tape your pencil or pen to a piece of string attached to your clipboard or binder so that it is there when you need it.
3. If your records are likely to get wet or dirty, consider laminating them and writing in permanent marker.
4. Make sure your forms are simple enough for anyone to use. If a column heading says '#' it should be obvious what number needs to be recorded.

Tips for PDAs

PDA's (Personal Digital Assistants) are hand-held, electronic notepads that allow you to record your information in the field, barn or processing facility. The information can then be directly uploaded onto a computer at the farmhouse without the need for transferring data by hand.

PDAs can also send information via the Internet.

» Carry your PDA in a pocket that has a button or zip, wear it on a string around your neck or attach it by a spring or bungee holder to your belt.

» This may seem obvious but, remember to bring your PDA with you when you go into the field. Keep it by the tractor key, on a shelf right by the door or inside your hat or boot.

» Purchase your PDA in a rugged, waterproof format so that it has a better chance of surviving if it is dropped in a manure pile or stepped on by a cow. If you already own a regular PDA you should consider buying a rigid and/or waterproof case for it.

» Look into available software programs for farmers. These include FarmNotebook.com at the low end and higher end products such as ProducePak, AgExpert Analyst and Farm Wizard. However, a simple spreadsheet will often do the job.

OUTPUTS:
CROPS, MEAT, EGGS, DAIRY, VALUE-ADDED PRODUCTS

Records of your production, sales and inventory in storage must be provided on an annual basis to your certifier. Since most of the information about your "outputs" is numerical, these records normally take the form of a paper chart or computer spreadsheet.

HARVEST VS. SALE QUANTITIES

Harvest, Processing and Sales data are not necessarily one and the same. For example:

PRODUCT	HARVEST AMOUNT: KG	PROCESSING: NO. OF 1 KG BUNCHES	SALES: NUMBER OF BUNCHES
Beets	78	75	74
Carrots	176	144	120

Virtually all beets were sold. Only a small number were lost due to damage or deformity, so the sales records approximate to the actual yield.

Many more carrots however, were rejected because they were forked or snapped. As a result, the number of bunches (the processing record) is quite different from the actual quantity of carrots harvested. Not all bunches of carrots were sold, so the sales number is smaller again.

It is important to keep harvest, processing *and* sales records in order to understand where losses occur and also so that you can improve your plans accordingly. The organic inspector meanwhile, needs to see what was actually grown, not just what was sold, to verify that the harvest and sales amounts are reasonable for the space and inputs used.

For products that carry labels, you'll need to keep samples on file to show organic claims (organic, organic content, organic ingredients, etc), the certifier's name or logo, ingredient lists and lot numbers. (The traceability section of chapter 4 deals with lot numbers in detail.)

CROPS EXAMPLE: HARVEST LOG

Harvest logs document not only the crop and the amount harvested, but also the date and field. Other relevant information, such as harvest conditions, may be recorded in the harvest log or in the journal. Harvest records need to connect easily to your field map and to inventory or sales records: use a lot number throughout that identifies the field that crop was harvested from. In field crop operations, the harvest log is often part of the field activities log (see the example in the Inputs section on page 31). In market gardens, the harvest record is generally a document of its own.

Blue Chicory Garden – Harvest Record sheet – by crop													Year: 2010	
				Quantities harvested and date										
Crop	Variety	Unit	Bed #	21-Jul	24-Jul	28-Jul	31-Jul	4-Aug	7-Aug	11-Aug	14-Aug	18-Aug	21-Aug	
Green Bean	Blue Lake	lb	T3&4						10	4	15.5	8	8	
Green Bean	Maxibel	lb	A3				3	3	3.5	5	3.5	1		
Yellow Bean	Rocdor	lb	C6	15	22	16	14	7	7	3				

LIVESTOCK & CROPS EXAMPLE: INVENTORY

Inventory records such as grain bin or freezer records are required for crops or products which are sold out of storage. This is necessary so that the organic inspector can perform an audit trail and so that you know how much you have available, whether for sale, as inputs for processed products, or as seed for the upcoming season.

PRODUIT	POIDS(g)	QUANTITÉ	TOTAL
Poitrine	300→309	1	
désossées	310→319	1	
X1	320→329	2	
	330→339	1	
	340→349	1	
	350→359	1	

Your inventory form should allow multiple in/out entries and cumulative totals. You may also include columns for lot numbers, suppliers, customers, and so forth.

For grain bin inventory (or whenever there is just one item per bin), simply update a form like the one below with any addition or removal.

BIN # 3 TYPE wood SIZE 1000 bushels *Northwinds Ninety*

DATE	PRODUCT & FIELD	OPENING INVENTORY	EST BUS. IN/OUT	ENDING INVENTORY	BUYER	INVOICE & LOT NO.	PRICE/BUS	TOTAL
96/09/12	RYE - 2	0	900	900		9623R		
97/10/10		900	(150)	750	used for seed for field 9			
98/04/20		750	(150)	600	used for seed for field 5,6 & 7			
98/04/27		600 est.	363.27 / 114.73 (screenings)	150	Reid Hill Farms	643561/9623R	363.27 bus @ [illegible]	2215.95
98/05/30		150	<150>	0	trans to bin 6			
98/08/23	8 oats		350	350		98833		
99/01/06	oats	350	<350>	0	Cleaned + moved to bin 7			
99/05/17	Flax seed	14 bags		14 bags				
99/10/02	Flax seed	14 bags	(14 bags)	0	moved to bin #6.			
99/10/9	Flax	0	200 bus	200 bus.		5272399		
99/10/18	Flax	200	<200>	0	to Morinville for cleaning + bagging then to Bin 6			
01/05/18	Rye seed	0	274.6	274.6.	Cleaned @ Westlock Plant not organic.			
01/05/01	Rye Seed	274.6	188	86.6	E. Keberenick	109879	5.25	987.00 conv.
01/05/07	Rye Seed	86.6	20	66.6	Robins Organics	109881	5.25	105.00 conv.
01/05/09	Rye Seed	66.6	298.6	365.2				
01/06/11	Rye Seed	365.2	196.8	168.4	Sold to Westlock Seeds	Sold as conv. seed		
01/06/14	Rye Seed	168.4	168.4	0	sold to Wade Meakin as conv. seed			
01/08/15	Rye Seed		75	75	purchased from Tony Heuks			
30/08/03	Rye	75	75	0	seeded field 14			
30/09/03	Rye	0	250	250				

FIVE TIPS FOR OUTPUT RECORDS

1. If possible, enter data directly onto a PDA or computer spreadsheet so you don't have to transfer the data later.
2. If you do need to transfer data, try to do it on a regular basis; make sure all of the original forms are stored in one location where they cannot get lost, have coffee spilled on them or get eaten by the dog.
3. Record sales as a weight, volume or quantity in addition to as a dollar value.
4. Keep photocopies of sales receipts if they need to be with the accountant at the time of organic inspection.
5. Tracking sales at the Farmers' market can be tricky (and computerized cash registers are expensive), but it's fairly simple to record the quantity at the start and end of the market and calculate how much of a particular product was sold.

Supporting documents: neighbours, suppliers

Written communications to or from your suppliers, neighbours and customers provide additional information to support your record keeping. For ease of retrieval they should be filed with or near the records they support.

Situation	Supporting documents
Using off-farm storage or transport	Affidavit from owner stating that the equipment has either only ever been used for organic production or has been cleaned thoroughly
Renting land	Signed letter from the owner stating what occurred on the property while under their management and the date of application of the last prohibited input
Purchasing uncertified seed, feed, straw, etc	Signed letter describing the growing practices used to produce the seed, straw, etc
Slaughtering	Documentation from the abattoir or butcher that explains how commingling and contamination are avoided

Affidavit and letter templates are often available from your certifier.

Letters may also be required for inputs without a full ingredient list, which may carry a risk of contamination or genetically engineered (GE) content. For example:

Product	Documentation required
Packaging and containers	Letter from the manufacturer stating that the product is food grade
Biodegradable products made from corn (if not organic, corn could be a genetically engineered variety)	Non-GE affidavit from the manufacturer

Product	Documentation required
Non-organic seed that has a GE equivalent	Non-GE affidavit from the seed grower or supplier
Brand name fertility or pest control product claiming to be organic, but not yet recognised by the certifier	Letter from the manufacturer listing the ingredients (including fillers and non-active ingredients) and responding to any specific issues in the PSL such as the use of ethoxyquin in fish fertilizers

Gord grows certified organic lentils for export. For every sale, he lists the sale in his sales control register and keeps the following supporting documents filed together. He records the lot number for each shipment on all documents.

- a copy of the sales invoice sent to the buyer
- a record of where it was cleaned
- a copy of the trucker's Bill of Lading (if applicable)
- a copy of his Bill of Lading
- a copy of the weigh scale ticket
- a copy of the Transaction Certificate Application (TCA)
- a copy of the Transaction Certificate (after it is received from his certifier)

BARNYARD ORGANICS

MARK & SALLY BERNARD, MARK'S FATHER WENDELL BERNARD

FREETOWN, PEI (NEAR SUMMERSIDE) WWW.BARNYARDORGANICS.CA

FARM:
Field Crops,
550 acres (223 ha),
270 acres(109 ha) in certified organic production

ORGANIC CERTIFICATION:
Atlantic Certified Organic Coop Ltd. (ACO)

FIRST CERTIFIED:
2006

RECORD KEEPING SYSTEM: Computer

"Before 1985, we were a conventional dairy and potato farm," says Mark. "We sold the dairy cows in 1985, continued with potatoes, then looked at the books and markets in 2001 and decided to take a couple of years off. I took a job off the farm but didn't like it, so I decided to go to the Nova Scotia Agricultural College (NSAC) for a business and plant science diploma, focusing on the business opportunities in farming."

"For new farmers, it's a good idea to find a niche market. Try to find a need. Fulfill it and if possible expand on it. We're meeting a growing demand for organic dairy, meat and poultry on the island by growing soybeans and cereal grains for livestock feed, with forage in the rotation. Through networking, I have met livestock farmers who prefer to buy soybeans rather than grow their own, so I sell all the soybeans and cereal grains we grow direct to those farmers. Sales have been so strong that one winter I forgot to keep enough for our own sheep and chickens. To deactivate the trypsin inhibitor and urease enzymes that cattle and other livestock can't break down, we bought a used soybean roaster.

"The farm has been in transition since 2003, with the first 45 acres certified in 2006, 120 acres in 2007, and a total of 270 acres in 2008. The rest will come on as we build the soil and work towards a five-year rotation. Transitioning is a learning process, so start slow. Don't transition all at once. Grow a bit each year as you learn.

"I won an award from Farm Credit Canada (FCC) for the business plan I produced at NSAC. The prize was a copy of their accounting software program, AgExpert Analyst. I also found FCC's Field Manager program on their website and started to explore that as well. I've used both from the beginning. I also downloaded templates from the ATTRA website for records that the Field Manager program doesn't produce. I set up those templates in a spreadsheet.

"Every year, the software updates contain something I've been looking for that allows me to reduce the number of separate spreadsheets that I have to keep. One recent update allowed me to track lot numbers and certificate numbers for seed purchases. Another included an automatic prompt: if you record an operation, you can see whether you want to record a cleanout affidavit for the equipment or storage container that you're using.

"To use the program, the first thing I do is define my farm, identifying each field as the smallest area for how it's used. For instance, a large field that is part soybeans and part oats is set up as two fields in the software. The next thing you do is list equipment, seed, labour, and other inputs – everything that goes on in the farm. Then, for any field work, all you have to do for the record keeping is choose that field, and the equipment and input used. You can also make additional notes about what seed you're planting or what equipment you're using.

"We have a little handheld computer (PDA) that travels with us in the tractor cab. If we're out planting a field and run out of seed part way around the field, we note the location on the PDA. Later in the day we plug it into the docking station and update the records on the main computer.

"If you use the software to track your costs, you can produce budgets and end-year cost of production reports. It merges with the accounting program for financial analysis.

"I keep just a few records on spreadsheets or in paper format. I keep my seed sourcing records and composting records in spreadsheets and print out only the reports I need. I do as little as possible on paper. Sally's records for the sheep are on paper only; for example, notes about breeding, and health treatments such as copper for foot-rot. The Field Manager software isn't designed for livestock.

"For a market gardener, the program could be set up by defining each field as an area used for a particular crop. There's no limit to the number of fields you can define. It could be worthwhile to figure out how to adapt it for a market garden, but the annual fee for the Field Manager program may be hard to justify for a smaller operation.

"Computer record keeping saves time compared to paper records. I like that I can access the data to show the whole farm or separate fields or each crop; for example, what date the soybeans were planted or harvested. Any of the organic inspector's questions can be answered when I bring up the relevant data on the computer. Or I can print it off in minutes, instead of digging for a half-hour through file folders. A couple of clicks and my information is right there.

"FCC is promoting the software to organic farmers. They want to make it work for us and so they consider our needs when planning updates.

"Even for conventional farmers, record keeping has evolved. For example, it's now mandated to keep spray records. Organic farming involves a little more record keeping, but conventional is moving that way. Organic record keeping means writing down everything that's being done on the farm so you can go back to it later. It may require a little more time, but it's not overly complicated."

Donalda Springs Organic

Rod and Cheryl Randen, Donalda, Alberta www.barnyardorganics.ca

Farm:
Livestock
2 1/2 sections
(1600 acres/648 hectares)

Organic certification:
QMI

First certified:
1999

Record keeping system:
Paper and Computer

Rod's grandfather settled the farm in 1902 when Alberta was encouraging homesteaders, and the operation passed from father to son, with Rod taking over in 1980. Chemicals have never been used on the land. The Randens have over 300 bison in a breeding herd raised on one section of native pasture. They also have one and a half sections in hay for winter feed and bedding.

Donalda Springs Organic has both Canadian and US sales. Rod credits his certifier for help with forms, advice, and standards information and for helping to keep his record keeping straightforward so that he can avoid any time-consuming mistakes. The certifier provided templates for the farm plan, field history, activity log, test results, and inputs log. Rod has customized some of these, especially the activity log, to better suit his operation.

For his field map, Rod used a satellite image and added details of his boundaries and buildings. Cheryl created a paper-based sales register for meat sales. They both record daily activities in workbooks kept in the tractor and truck. When inspection approaches, they review all documents for completeness and accuracy. "The more informative the records are, the easier it is for the organic inspector," says Cheryl. Rod admits to "too many late nights" as the downside of preparing the required documents, but cautions farmers to "remember you are paying by the hour, so it's worth your time to have all documents and receipts ready for inspection."

He suggests that new farmers learn from experienced organic operators in the area. "Ask for help with what works or doesn't work in your specific conditions," he says. He also advises farmers seeking organic certification for the first time to use a consulting firm to help prepare the initial application. "It's really important to get started right."

3 APPLYING TO CERTIFY – THE ORGANIC PLAN

Your organic plan is a description of how you do your work. It consists of procedures and possible strategies. It's a quality system for your operation.

Monique Scholz, organic inspector, QC

AN ORGANIC PLAN is essentially a set of **procedures** that describe how you operate your farm; **records** are the finishing step, showing how you implemented your procedures.

Each certifier has its own application form. It contains questions that help you describe your operation and practices, guiding you, in effect, in preparing an organic plan. In addition, your certifier may provide other forms if your operation includes greenhouse or wild crafted crops, maple syrup, honey, mushrooms, on-farm processing, or other specialties. After you apply the first time, you only have to update your original plan in subsequent years.

The application form has all the elements of a basic organic plan. Some farmers however prefer something that is more in-depth and provides detailed descriptions of their production methods and practices. In that case, the organic plan becomes part of their business plan or procedures manual.

"In-depth" doesn't necessarily mean writing volumes however; your organic plan can be complete in four pages, or it can be longer. The detail and length depends on what is useful for your operation.

Initially, when you set up your organic plan you may be left with as many questions as answers. That's okay. The purpose of an organic plan is to encourage you to think about your operation, plan for most eventualities and, of course, read and comply with the Standards.

Your certifier knows that over the course of the year your organic plan will change. It's not supposed to be static. If there are significant changes from the time of your application to the time of your inspection, you should submit a revised plan to the organic inspector. Other changes will be documented in your next application/update.

This chapter provides examples and tips on how to make the organic plan effective for you, your certifier and the organic inspector.

ORGANIZING YOUR ORGANIC PLAN

Whether your operation includes crops, livestock, on-farm processing or a combination, your organic plan will describe:

- your farm – including a Farm Map and Field History
- your inputs – everything that you plan to bring in to the farm to grow or to promote growth and health
- your activities – your growing and management practices
- your outputs – how your final products (crops, eggs, dairy, meat, value-added items, etc) are handled, stored, transported and sold

While "activities" and "outputs" may be unfamiliar terms, it's helpful to think of your planning and your subsequent record keeping as three stages. Thinking in terms of inputs, activities, and outputs gives us a way of subdividing what could be an overwhelming list of requirements.

All organic plans have the same basic structure. The Records Master Chart (Appendix A) lists the information required in each section. Keep this outline handy as you read the instructions and examples that follow.

Your Certifier's application may organize the information presented here differently, or ask for more or less detail.

You will also need the latest Canadian Organic Standards CAN/CGSB 32.310 General Principles and Management Standards and 32.311 Permitted Substances List (PSL)[9].

FARM MAP AND HISTORY

Maps and charts in this section of your organic plan show your:

- farm location
- farm layout
- field history
- rotation plans

FARM LOCATION

The purpose of the farm location map is to make it as easy as possible for the organic inspector to find your farm. Print a map off the Internet from Mapquest™ or Google™ maps, or copy a road map that shows enough detail

9 Or your national Standards. The COS is available from the Canadian General Standards Board on the web at www.tpsgc-pwgsc.gc.ca/cgsb/ or by calling 800-665-CGSB.

to locate your farm from the nearest major town. Highlight the best route, or draw a freehand, legible route to your farm. Print your name, your farm name, address (including pin or fire number) and phone number(s). Including a photo of your farm's road-sign or turn-off is a nice touch.

Always assume that your organic inspector does not know the area, and avoid using landmarks that are no longer there. Most inspectors have funny but frustrating stories about trying to find the turn-off where the old burned down church used to be.

"I arrived at one farm two hours late, totally frazzled and covered in mud after first getting stuck in a bog after trying to drive down an old logging trail, then nearly driving into an abandoned quarry with a lake at the bottom. The inspection went very well in the end, but it was not an auspicious start."

- anonymous Organic Inspector

FARM LAYOUT

Your organic plan must include a bird's-eye view of your operation, showing the locations of:

- fields (with names or numbers)
- boundaries, buffer zones and neighbouring land use
- buildings: farm, storage facilities, processing facilities, animal accommodations, etc; each labelled or numbered
- water (ponds or watercourses)
- roads
- other infrastructure or physical features, including manure piles and tanks, septic fields, other sources of potential contamination
- any nearby risks such as conventional farms or industries

This can be done with maps, aerial photographs, or hand-drawn diagrams identifying the features listed above.

Use an appropriate scale. A field crop or livestock operation with several sections of land or rented farms may need one map as an overview and separate maps for more detail; a small market garden may need only one sheet. Be sure to mark the orientation and scale on the map, as well as the prevailing wind direction.

Maps can be one of several types:

- topographical maps (photocopied and enlarged if necessary)
- survey maps
- aerial maps from municipal or provincial offices (there may be a fee)
- online maps: you may be able to find a satellite view with enough detail on Google™ maps or Google™ Earth you if you're lucky or if your farm is very large

Use standard size paper for photocopying. Maps must be drawn to scale and be an accurate representation of the farm. If you use a topographical map be sure that the resolution provides enough detail and that the format allows you to add other information; otherwise a hand-drawn map may be easier to understand.

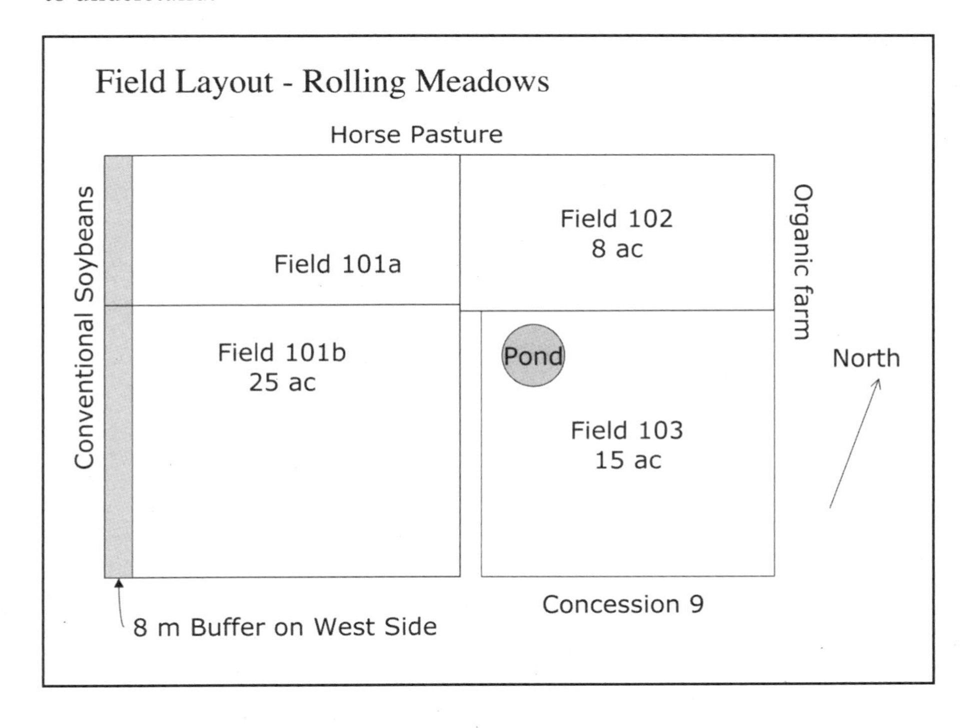

TEN TIPS FOR BETTER MAPS

1. For a hand-drawn map, use a ruler if the edges of your field are straight.
2. Add numbers and a legend if your map is cluttered with details.
3. Use a key or add dates to crop labels if you practise succession planting.

D1 Amish Tomatoes, 10 Jun	D2 Scotia Tomatoes, 10 Jun	D3 Scotia Tomatoes, 10 Jun
C1 Harris Parsnips, 20 Apr	C2 Harris Parsnips, 20 Apr	C3 Harris Parsnips, 20 Apr
B1 WK* Cabbage, 22 Apr	B2 WK Cabbage, 22 Apr	B3 WK Cabbage, 22 Apr
A1	A2	A3

Bed	Succession Plantings
A1	Arugula April 23rd to May 30th. Mixed lettuce June 1st to July 15th. Golden Detroit Beets July 20th to September 30th.
A2	Garlic Overwintered to August 2nd. Sugar snap peas August 4th to October 30th.
A3	Carrots April 30 to July 30th. Oats and Crimson Clover August 1st to October 20th (tilled under). Garlic October 28th overwintered.
B	WK = Winterkeeper Cabbage

Using a key to show succession plantings in bed A

4. Use the same names and numbers that you normally use when labelling fields and beds. If you call a field 'North Field' then label your map 'North Field'. Don't rename all your fields just for certification purposes.
5. Use a naming system that's easy to remember.
6. Be consistent in your numbering; e.g. Field 1 is the closest to the Farmhouse, while Field 9 is furthest away, or Field 1 was purchased first while Field 5 was acquired more recently.
7. If your operation spans more than one farm, consider numbering them in blocks: the first farm is 100, with fields 101, 102, and 103; the second farm is 200 with fields 201, 202, 203, etc.
8. If you divide a field into two, name the subdivisions after the original field with the suffixes A, B etc. For example, Field 3 will become fields 3A and 3B.
9. If you have more than one row of beds consider using a grid so that 6B is the 6th bed along, two beds deep.
10. To rename combined fields use the original numbers or the next available number. For example, if a farmer owns 8 fields and Fields 2 and 3 are combined, they should be renamed 'Field 2/3' or 'Field 9'.

FIELD HISTORY

Your Field History and Farm Journal are your two most important documents.

Your Field History includes:

- the size of the fields (in acres, hectares, square feet or square metres)
- crop or grazing rotations
- applications of any soil amendments (compost, manure, lime, rock phosphate, etc)

Your annual application includes a Field History chart. It summarizes the crops and substances applied to the soil over several years (at least three years past plus the current and upcoming year). Your field history contains a lot of important information on crop rotations, nutrient accounting, green manure, cover crops and inputs, including inoculants (keep labels). Be sure to include dates and amounts of anything applied to a field, including manure. Other information such as the source of the input and the application rate would be recorded in more detail in your journal or on the field activity log (see chapter 2).

Field History - Rolling Meadows

Legal Desc. (Address or MPAC number)	Field No.	Field Area (ac)	Last Month & Year of Unpermitted Substance Use		2007	2008	2009	2010
187 Conc 9 Conc 9E Pt3 Lot2	101	25	May, 2009	Crop Fertilizer Pesticide Other	Corn **Urea** **Furadan** -	Soybean - **Matador** -	Corn **Urea** - -	Soybeans - - -
187 Conc 9	102	8	July, 2006	Crop Fertilizer Pesticide Other	Pasture - - -	Pasture - - -	Pasture - - -	Pasture - - Overseeded
187 Conc 9	103	15	May 2006	Crop Fertilizer Pesticide Other	Hay Manure - -	Hay - - -	Hay Manure - -	Hay Reseeded Manure - -

ROTATION PLANS

Organic methods generally involve long and varied crop rotations. A good crop rotation includes legumes and green manures (plowdowns) to build soil, organic matter and biological activity. The organic inspector will check that your rotation plan is effective in managing weed, pest, disease and fertility issues. The simplest format for a rotation plan is a list. If you vary rotations on different parts of the farm (one rotation for pasture and another for crops, for example), indicate the fields to which each plan applies.

Your Field History includes plans for the upcoming season, but your rotation plan should extend at least five years into the future; some even suggest up to ten years.

FIVE TIPS FOR BETTER FIELD HISTORIES AND CROP ROTATION PLANS

1. Use the same numbering or naming system on your map and in your field history.
2. Record your field histories in a spreadsheet on the computer so that you can make changes easily.
3. If you do your field history on paper, pencil in your plans for the current year and ink them in as they occur (or erase and correct them if they don't).
4. Highlight fields that are in transition and make sure that you have a field history that covers the entire transition period.
5. If you practise succession planting, make sure to list ALL crops grown – including green manures and plowdowns – with their dates.

INPUTS:
SOIL AND AMENDMENTS, TREATMENTS, SEED, LIVESTOCK, PROCESSING INGREDIENTS, CLEANERS

This part of your organic plan covers everything that you intend to use on your farm, including seed, feed, soil amendments, manure and compost, purchased livestock, organic pesticides, biodegradable mulch and cleaning products.

Make a list of all inputs, grouped by the subheadings in the Records Master Chart (appendix A), then add how they will be used, why they were chosen (and whether there are any alternatives), plus supporting evidence from the supplier to demonstrate their compliance with the Standards, such as:

- MSDS or ingredient list
- GE-free statement
- organic certificate

Here's how this might appear in your plan:

PRODUCT (SUPPLIER)	USE
DiPel	Control of cabbage worms. Monitor for white butterflies and apply when they are first observed. Preferred method of control because floating row cover always blew off and trap crop strategy did not provide adequate control. Letter from certifier on file.
Fish Fertilizer	Used in soil block mix as per directions on bottle. Does not contain ethoxyquin (letter on file).
Hydrogen Peroxide	Food grade, 35%. Added to drinking water of chickens at a concentration of 40-80ml in 10 litres of water
Garlic Powder	Added to sour cream dip. Certified organic by OCIA.

GENERIC INPUTS

Check the Permitted Substances Lists (PSL) of the Canadian Organic Standards to find out which generic inputs are allowable and under what circumstances. Note whether the product is permitted or restricted in its use. Make sure that you are looking in the correct section of the PSL, which is divided as follows:

Crops

- soil amendments
- crop production aids and materials
- weed management

Livestock

- feed, feed additives and feed supplements
- health care products and production aids

Processing

- food additives
- processing aids
- pest control substances

Cleaning

- cleaners and disinfectants

If you can't find the input in the PSL, do some research (online or with your certifier) to find alternate names under which it might be listed. If you still can't find it in the PSL, then it is not allowed for use under the COS and you'll need to find an alternative.

When sourcing a generic input such as lime or manure, record any research that you do to ensure there has been no contamination of the product during mining, processing or transportation. COG's Guide to Understanding the Canadian Organic Standards provides more information about what to look for and questions to ask suppliers.

MANURE AND COMPOST

Your plan shows your intended sources of compost and/or manure. These sources must comply with the Standards (see page 15). If any ingredients or additives are prohibited or restricted, your organic plan must include an appropriate composting process for the manure. For instance, certain medications such as sulfamethazine do not break down effectively as a result of composting, so you must account for this in your plan. Restricted substances must be certifier-approved before use.

SEED SEARCH

Certified organic growers must plant certified organic seed where it is commercially available. This applies to all seed, including that of green manures.

Growers may purchase uncertified organically grown or conventionally grown seed only if they can demonstrate that no organic seed was commercially available. Neither genetically engineered nor treated seeds are acceptable. If your organic plan indicates that you do not intend to purchase certified organic seed when it is apparently available, then you need to explain why. Price and lack of organization are never acceptable rationale.

If the inputs list in your organic plan included certified organic seed sources, but subsequently you could not source organic seed, then you must complete a seed search document to demonstrate that you tried to source certified organic seed (see example in chapter 2).

You must contact your certifier for permission before planting a non-organic substitute.

BRAND NAME PRODUCTS

For brand name products containing inert materials, fillers, and non-active ingredients not listed on the label, you should first check with your certifier prior to using a new product; many maintain their own brand-name directories. If requested by the certifier, you may need to ask the manufacturer for an ingredient list or MSDS so that it can be assessed for compliance with the PSL. In some cases, you will have to get a non-GE affidavit. Companies that sell primarily to certified organic farmers may have a standard letter on file for this purpose. Keep the ingredient list, the letter, or the MSDS on file for inspection.

VACCINES AND ANTIBIOTICS

Vaccines and other medical treatments may contain antibiotics or GE ingredients. Contact your certifier prior to using any product that they have not previously approved.

Your organic plan must specify how you will track livestock that have received antibiotics (to ensure they are sold as non-organic) and what happens to manure from antibiotic-treated animals.

PRE-MIXES

Some livestock mineral pre-mixes contain prohibited medications or preservatives. You must show in your organic plan that you will be sourcing pre-mixes that are free of these prohibited ingredients.

BIODEGRADABLE PLASTICS

If you buy products such as corn-based biodegradable plastic packaging or biodegradable mulch, you must document that the corn is not genetically engineered. Some products include plastic binders that do not degrade. If plastic is not "fully" biodegradable you must spell out how you will dispose of the product at the end of the season.

SIX TIPS FOR INPUTS IN THE ORGANIC PLAN

1. Make a note of any information that you will need to keep on record and the format that the information will take: labels, ingredient lists, receipts, letters or forms, for example.
2. Set up forms or storage files for your input records. For example, for receipts, have separate file folders ready for seeds, supplies, containers, feed, building repairs, etc. (Extra tip: use the categories that you will need for your tax return, and you'll be way ahead of the game – just pull out the seed and feed folders at organic inspection time). Put receipts in the appropriate folder as you get them.

3. Use restricted inputs according to the Standards. For example adding boron as a soil amendment requires prior evidence (soil test) of deficiency.
4. Be especially vigilant for ingredients that may be genetically engineered: corn, cotton seed, soy, canola, sugar beets and papaya and their derivatives, such as xanthan gum and dextrin. Buy certified organic inputs where possible or get non-GE affidavits.
5. Be aware that products that are OMRI-listed are not necessarily compliant with the Canadian Organic Standards. Check with your certifier to see if they have a brand names list.
6. Be aware also that some commonly used products that conform to the COS have not been approved for use in Canada by the PMRA. These products cannot be used by ANY farmers in Canada.

ACTIVITIES:
FIELD WORK, HEALTH MANAGEMENT, ON-FARM PROCESSING

"Activities" is a catch-all term used in this handbook to include the parts of your organic plan where you describe the processes and systems that you use in your operation. Everything that's between "inputs" (the seed, animals, ingredients, or soil amendments that you start with) and "outputs" (the crops, animal products or value-added products that you end up with) – that's where your production activities are.

The Records Master Chart (Appendix A) is organized under the headings of Crops, Livestock, and On-farm processing. Use the subheadings to help guide you in choosing which sections of the organic plan to complete for your operation. For instance, some of the subheadings under Crops include management plans for weeds, pests and diseases. If you are using a certifier's application form as a guide, you will see roughly the same headings and subheadings.

Under each subheading enter the following information:

- production practices and procedures, referring to any specific inputs that you normally use (as listed in the Inputs section above)
- management practices and physical barriers used to prevent contamination and commingling
- potential risks and solutions (both known and untried)
- monitoring and record keeping that you plan to do

Your organic plan and your subsequent record keeping work together. Your plan lays out your normal and expected processes, and your records document what actually happens. In addition to showing how your activities comply with the Standards, you will be able to assess how well your processes work and whether you need to make changes for the following year's update. Here is an example that shows the continuity between your organic plan and your records:

PEST MANAGEMENT

In your **organic plan**, you would include a line under the Pest Management subheading for your normal strategy to control cabbage pests:

Cabbage family – trying companion planting with marigolds this year; will use Dipel if pest levels warrant it

Also, in the Inputs section you would list Dipel and its manufacturer.

In your **records**, you would have on file:

- Product label from Dipel showing ingredients
- Letter from Safer (manufacturer) describing the inert ingredients
- Farm Journal entries such as:
  – June 12: Cabbage loopers noted on broccoli. Companion planting doesn't seem to have been effective. Dipel applied per label.
  – June 26: 2nd application of Dipel on broccoli.
  – July 10: no further evidence of looper damage.

When you update your plan for the following year's application, you would probably drop the companion planting idea, or indicate a new strategy that you will try.

Following are a few examples of how to document an activity. There is no set style for completing this section of your organic plan– essay, lists, tables, charts, a question/ answer format–whatever suits you and your certifier. The examples below tend to be longer than many certifiers require. The detail included here is for clarity. On your actual application however, it's good to be concise. Just be sure you demonstrate awareness of the Standards when describing your processes.

LIVESTOCK EXAMPLE – DISBUDDING GOATS

Sean raises goats for meat and milk. One of the regular tasks is disbudding kid horns. His organic plan looks like this:

Process
Goats are disbudded at 5 to 10 days of age. This is done using a disbudding iron applied for 10 seconds to the horn buds to burn away all horn-producing cells.
Records
Date and names or ear tag numbers of the goats are recorded in the Farm Journal.
Management practices and physical barriers used to prevent contamination and commingling
n/a
Process Description
Dehorn kids when you can feel the point of the horn pushing through the hair. As each horn bud is done, check the kid's condition to make sure it is not suffering from shock. Kid should be conscious, moving and responding to touch. After disbudding, use antiseptic spray on disbudded area and bring kid to mother. Kid should suckle immediately (if kids aren't on does, give them a bottle immediately after disbudding). Watch kids for 20 minutes or so after returning them to their mother (or to the pen of kids) to make sure they are moving around and not showing signs of shock (lying limp in a corner). If kids are looking lively and active, they should be fine. Check disbudded area daily as scab forms to make sure there is no bleeding or infection.
Potential risks and solutions for Livestock Health
Risks: animals going into shock during disbudding.
Solutions: disbud goats as young as possible, as quickly as possible and put the kids back in with other goats (or mother) as soon after the operation as possible. Don't hold the iron to the head for more than 10 seconds at one time to avoid burning too deeply, which can cause brain damage. Pause and let the area cool for a few seconds if you have to do it again.
Potential risks and solutions for certification
Risk: Prohibited substances in antiseptic spray
Solution: letter from manufacturer
Risk: Infection, requiring treatment with antibiotics.
Solution: Take care to keep all equipment clean, follow procedures with care and monitor carefully after the fact.

Although the "process description" above is not required for certification purposes, there is no harm in including more detail. If such additional information serves your management needs (for training or tracking research, for instance), then include it as you see fit.

Note that in this example there are two kinds of potential risks: those that could impact the health of the animals and those that could impinge on organic certification. It makes more sense to have one document containing all of this information than to maintain one for certification purposes and one for livestock management. Planning for both kinds of risks makes for better farm management and draws the attention of the organic inspector and the certifier to areas where more information may be required; for example, does the antiseptic spray contain only permitted substances?

CROP EXAMPLE – STRIPED CUCUMBER BEETLE CONTROL

Bruce and Hanna have a crop binder where they keep notes on each of the vegetable varieties in their market garden. When preparing their organic plan, they extract pest control strategies such as the following:

Crop	Pest	Control
Squash	Striped Cucumber Beetle	Row cover on by June 1st. Bug suckers or propane flamers as back up. Monitor the arrival dates of the cucumber beetles and their numbers using sticky traps and record this information on a calendar. (We use the same calendar to record the arrivals of all problem insects and diseases.)

Note that since there is no obvious risk of contamination or commingling in this situation, there is no mention of it under "control."

PROCESSING EXAMPLE – SANITIZING MEAT CUTTING BLADES

Zak raises specialty meat (fallow deer, rabbits) and has on-site inspected processing facilities for meat products (the animals are slaughtered at a certified abattoir). However, his example applies also to farmers who market their own meat that is processed by a butcher; the farmer needs to oversee the process to ensure that organic standards are met.

Cleaning process: After cutting each batch of meat, the blades on all meat cutting equipment must be sanitized following these steps:

» Put on rubber gloves
» Prepare a standard bleach solution in a well-ventilated area
» Soak blades for 5 minutes
» Scrub to remove any residue

» Rinse twice to dilute the wash water to tap water concentration
Rinsing prevents contamination of the meat by the sanitizer. Rinse water is disposed of down the drain. Each sanitizing event is recorded in the cutting equipment sanitizing form (see below), which has these instructions written at the top.

CUTTING EQUIPMENT SANITIZING FORM

» Put on rubber gloves.
» Prepare a standard bleach solution in a well ventilated area.
» Soak blades for 5 minutes.
» Scrub to remove any residue.
» Rinse twice to dilute the wash water to tap water concentration.

Date	Personnel		Date	Personnel
02/05/09	Mike Smith			
03/05/09	Julie LeBlanc			

FIVE TIPS FOR ACTIVITIES IN THE ORGANIC PLAN

1. List the names of inputs used in any activity. No need to bother with input sources or organic status; this information is already in the Inputs section.
2. Describe any standard process that will be carried out multiple times; for instance, feed rations and regular health care procedures, or compost tea preparation if you always use the same recipe. By describing it in your plan, you don't have to repeat the full description in your journal each time it occurs.
3. Use the certifier's application forms as a guide for deciding which processes to describe in your plan for certification purposes. Record any others in your farm's procedures manual.
4. Think about what you're going to need to document and what your daily records will need to look like. Use a template from the certifier or other source, or design your own log.
5. Laminate a copy of any instructions (ideally, formatted as a check-list) and keep it with the log for that activity. For example, keep the truck clean-out instructions with the truck clean-out log in the vehicle shed.

OUTPUTS:
CROPS, MEAT, EGGS, DAIRY, VALUE-ADDED PRODUCTS

This section of your organic plan describes all of the things that you produce on your farm plus what you do with them:

- describe your normal harvest, handling, storage, transportation and identification procedures as well as your sales or disposal practices
- identify points where there is a risk of contamination and commingling and your methods for reducing those risks
- list the types of records that you will keep at each stage of your operation
- outline your identification system for livestock, crop storage bins, or product lots and describe how they can be traced within your system
- include predicted yields and sales

Again, check the master chart in Appendix A for a full list of headings for your output records and plans. The following are some examples from organic plans for crop, livestock and on-farm processing operations.

CROP EXAMPLE – FLAX SEED

Steve has been growing flax for several years and recently transitioned to organic methods. He describes his normal practices under the appropriate headings on his certifier's application form:

Harvest:

Use our own combine harvester around mid-September when moisture content is roughly <12%, preferably < 10%. Record date and yield by field number on Field Activity Log.

Post Harvest Handling & Storage:

Dry the seed in our grain silos to <9.5% to 8% for long-term storage. Often it dries in the field to these levels. Assign lot # based on field #, bin #, year and certification #. Get genetic-engineering test done and save results on file.

Transport:

Transfer to a cleaning plant by bulk truck (get clean-truck affidavit) when ready to ship. Ship by end user or trader with Bill of Lading.

Documentation:

Documentation carries our farm name, product, organic logo and lot number on a Bill of Lading. We use Excel spreadsheets for bin records, sales and user fee records with lot and Transaction numbers. These are very basic but effective for updating year to year. (We have 34 fields, so we also do crop planning in a spreadsheet format. We can shift crop ideas from field to field to get the right balance of crops.)

LIVESTOCK EXAMPLE – FROZEN CHICKEN

Emelie answers questions frequently from customers wanting to know how her meat chickens were treated. She developed a fact-sheet for her employees and can extract from it for her certification application:

Chickens are gathered from the barn at 4 am having not received any feed for the previous 12 hours. They are placed into transport cages (5 per cage) and loaded onto a flatbed truck.

The birds are transported to the local provincially-inspected abattoir where they are the first birds to be processed in the morning after the equipment has been cleaned with PSL-compliant products; no other birds are on site at the time.

Depending on the demand, we ask for a mixture of whole chickens and parts: breasts, thighs, hearts, liver, necks, drumsticks, cubed and ground chicken. No spices, salt or pepper are added to any of the meat as our customers prefer unseasoned products.

The processed meat is stored in a designated 'organic' part of the abattoir and then collected first thing the following morning. All of the meat is wrapped, frozen and stored in boxes labelled with our farm name.

We only have organic meat on site, and we sort it out according to slaughter date (lot number) and cut of meat. Most meat is stored in our walk-in freezer, but meat that has been sorted for orders is stored in chest freezers in the shop.

All meat is labelled with the farm name, certifier, product description and date of processing, which we use as a lot number.

All of our meat products are sold through buying clubs. We list our products on a local food producers' website and our clients place orders which we receive directly through the website. We pack the orders into re-usable polystyrene boxes with cold packs and deliver them to central collection points once a week.

PROCESSING EXAMPLE – HONEY

Malcolm's apiary is in year 2 of a 4-year expansion plan. Next year when he acquires larger equipment, he will have to update the process descriptions in his organic plan, but while still dealing with relatively small quantities of honey, his extraction processing is described like this:

Harvest:

Honey is harvested twice per season, in early July and again at the end of August. However, this depends on colony health, weather and foraging conditions, and the amount of honey in the hives compared to colony needs. We harvest only from the half-size frames in the top supers (there may be 1-3 of these per hive), and not from the bottom two boxes.

Preparation of the Honey-processing room:

Before AND after an extraction, the following cleaning procedure is used and logged:

» wash down all counters, the extractor, pails and sieves with water and (biodegradable) dish-washing soap
» rinse all the above with water
» spray all surfaces with 35% hydrogen peroxide
» allow to air-dry before use

Processing:

Remove the frames from the hive, brush off the bees, place the frames in a storage bin and cover them. Take full bins to the honey-processing room. Cut the cappings off each frame with a knife, then put in the extractor and spin out. Pour out through largest-mesh sieve into a holding pail. Let settle; skim off any surface debris. Pour through finest-mesh sieve into sale containers (sanitized jars and/or pails; see separate procedure), cap tightly, and wipe off containers with warm water.

Grading:

Follow instructions in the refractometer case.

Labels:

Our pre-printed labels follow Ontario Regulation 384 requirements. At packing time, write net container weight on label in top right corner (250 ml = 330 g; 375 ml = 500 g; 750 ml = 1 kg).

Pack:

Put full, labelled honey containers back in the appropriate boxes by jar size. Write the lot number on the visible side of the box.

Records:

Each extraction is recorded in the Honey Log with date, total yield, package sizes and quantities, grading and observations.

FOUR TIPS FOR OUTPUTS IN THE ORGANIC PLAN

1. Harvest figures are not necessarily the same as sales figures, particularly when a product can be stored for a long period of time. If anticipated sales for grain in a given year are greater than expected yields, you will need to explain why (e.g. previous year's harvest is included in current year sales).
2. If you share equipment, transportation, processing or storage facilities you will need to explain how you will avoid commingling or contamination of your product.
3. Remember that labels and sales receipts must include the name or logo of the certifier.
4. Lot numbers must be used anywhere that a product is not being marketed directly to the consumer. They should be simple and consistently applied. (see Traceability in Chapter 4)

"Most people are already farming before they become certified. People farm by virtue of what they've learned from their parents, from extension agents, from equipment and supply dealers, from coffee shop conversations, from their own experience…

"Farming knowledge is very intuitive and unorganized, so it's not intuitive to sit down and plan on paper. The development of the organic plan has certain aspects that have to be written such as the field maps and field histories, but there are also aspects that change. For example, coming up to planting season, the weather prescribes that you do something different than what you thought you were going to do. So there's some fluidity that takes place. I do a lot of planning for crop rotations, new varieties, ways to avoid problems from previous years, and where the irrigation is going to go. It's not a formal kind of thing. Most parts of the plan evolve as you farm."

- Maureen Bostock, organic farmer and inspector, ON

Les jardins de la montagne

Joscelyne Charbonneau, Sylvain Brunet, Rougemont, QC

www.jardinsdelamontagne.com

Farm:
Market garden, 25 acres (10 ha)in production

Organic certification:
Quebec Vrai

First certified:
1997

Record keeping system:
Computer

Ten baskets for ten weeks – that was the beginning of Joscelyne's CSA in 1997. By 2008, Joscelyne, Sylvain, and ten employees were producing 516 baskets for 48 weeks of the year. And they still love the adventure of it.

With 53 species and 350 varieties grown each year in 25 acres of fields and two greenhouses, Joscelyne stresses that they "strive to minimize complexity, and simplify the process and time requirements for record keeping."

The website provides a lot of information to the public, and behind it is a farm-only view that provides a powerful set of tools to manage their CSA files and the daily operations of the farm. Sylvain and a friend wrote the management application themselves when he could not find suitable software ready-made. (CSA farmers interested in acquiring the software can contact Sylvain through his website.)

Here's how Sylvain describes their computer system: "We have computerized all of our accounting and record keeping. The sales register is part of the accounting system. Our website's "back-end" (visible only to the farm managers) includes the farm journal that tracks all daily activities. We try to avoid duplication of data entry and we are integrating and centralizing more of our record keeping into the Internet-based management system.

"If a document is too complicated to produce electronically (the farm map, for example), it is hand-produced, then scanned and kept in PDF file format. Paper forms such as soil test results, affidavits from seed suppliers, and purchase receipts are scanned in. All files are organized by year, and backups are done daily using the son-father-grandfather principle with off-site rotating copies.

"The seeding journal is updated in the greenhouses and scanned in. The transplanting journal is updated electronically each day by the farm manager via the website as is the daily activity journal. The irrigation journal is updated manually in the field when used, and scanned in an average of once a week."

Because of the number of varieties that they grow and the annual crop rotations, their certifier agreed that a field map was too complex. Instead, they create a hand-written field history (scanned into a PDF file) and make a visual record of crops for each section with a digital camcorder and digital camera.

To prepare for inspection, Sylvain and farm manager Enes review and complete the required documentation ahead of time, pulling together paper and computer records, and printing off the documents that the organic inspector will want to see and submit with her report. During the inspection they take notes. Upon receipt of the inspection reports, they correct any requirements and supply required paperwork to the certifier.

"At the beginning, with our lack of experience, we tended to be overzealous with record keeping – too much information and too much complexity," says Sylvain. "If you're new to this, ask around, copy and adapt what is already out there. You don't have to reinvent the wheel. You can save time and money by learning from others."

"It is challenging to make sure we collect and record everything that is needed for certification, and, just as important, useful for us from a management point of view. It is also challenging, but essential, to stay on top of things as much as possible. There are so many elements and so many variations that it can quickly become overwhelming and discouraging. We're only human and we then tend to skip doing it, but then we lose valuable information that could help in the management and better operation of our farm."

4 GETTING READY FOR INSPECTION

"I try to have people and records ready on time for the inspection, because I know from my own experience as an organic inspector that if the paperwork is ready, then the information that I need is available. And that makes the inspection go more smoothly and productively."
Roxanne Beavers, organic farmer and inspector, NS

IN THIS CHAPTER we'll focus on getting your records ready for the organic inspector's scrutiny. First we'll discuss how the inspection process impacts the way you keep your records and then what to expect from the inspection day itself.

KEEPING RECORDS FOR INSPECTION

The purpose of inspection is to verify your compliance with the Standards. It has two main parts: observing your risk management practices and verifying traceability. Keep these in mind as you create your organic plan and your records.

RISK MANAGEMENT

Once the certifier receives your completed application the organic inspector will read it with an eye to identifying "organic control points" or places or times when contamination or commingling with non-organic substances could occur. The organic inspector will evaluate your plan and your operation for evidence that you can mitigate the risk factors.

Examples of risk points:

ORGANIC CONTROL POINT	RISK	STRATEGY
Visually-identical organic and conventional crops stored in side-by-side bins	Easy to confuse organic and non-organic	Change from parallel production (not permitted under the COS) to all organic production
Spray drift from a neighbouring conventional farm	Contamination with herbicides	Buffer zones adequate width to protect the organic crop

Higher-than-normal parasite pressure on pasture caused by wet weather, and nothing in your plan on how to handle that situation without resorting to anthelmintics	Treatment may affect organic status	Fence 2 new pastures to increase grazing rotation
Unpermitted sanitizers in processing facilities	Contamination	Check with certifier for allowable brand-names
Bagging of organic and non-organic dried cranberries on the same production line.	Commingling	Careful use of cleanout process and log between organic and non-organic lots

As you create your organic plan and update it each year, look critically at your operation to identify possible risk points yourself, and include them in your plan, along with your strategies to overcome the risks. In some cases you can be proactive; for example, you could communicate with a conventional farm neighbour that you are certified organic and ask them to sign a letter confirming that they do not spray on windy days. In other cases you can document a "what-if" scenario, such as how you would handle aphids if you have an outbreak of them in your greenhouses in the coming season.

Any risk points revealed by a past inspection must be addressed in the next update to your organic plan. It may be appropriate to note the kind of monitoring you will be doing (and recording) to avoid or minimize the problem; for example, checking for evidence of herbicide drift into your buffer zones by making notes about the weed populations there.

TRACEABILITY

Your record keeping should be thorough enough to allow you to trace any of your products from purchase to sale. This is known as an 'audit trail.' It verifies:

- the completeness of your records
- the correlation between your production capacity and your product sales
- specific points of compliance (such as input sources being organic)

More examples of what the organic inspector is looking for during a traceability check are listed in the "Records Review" section of "The Inspection Process" below.

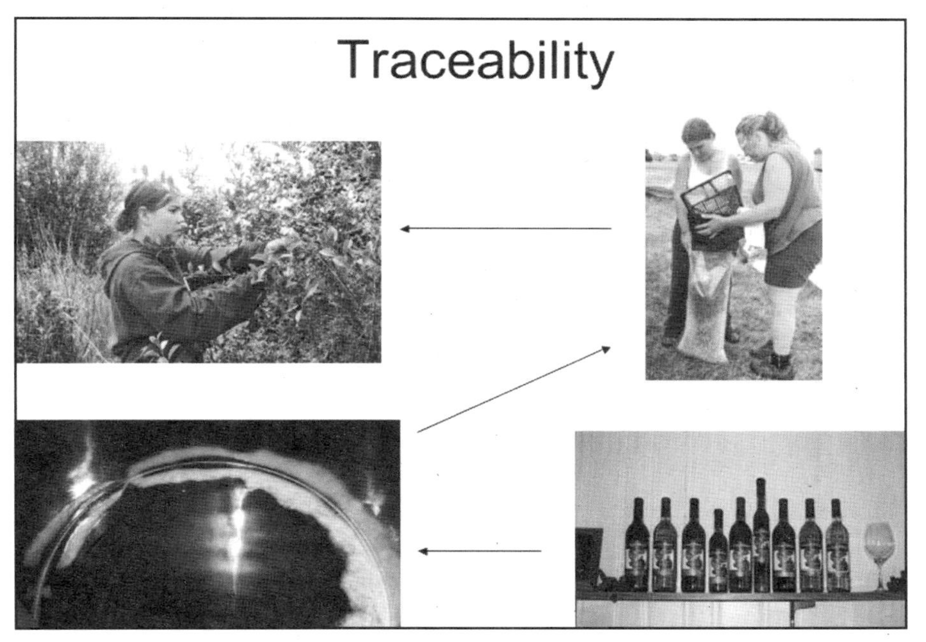

Traceability is valuable for farm management as well as for inspection; it ensures quality control, health and safety, and organic integrity.

QUALITY CONTROL

Imagine that a regular customer complimented you on a particularly great salad mix, jar of jam or batch of sausages. How would you be able to determine if you had done anything different with that batch?

In an ideal system you would be able to trace back to determine who made that batch, what ingredients were used, and where they came from. You would then be able to repeat the process to recreate your new and improved product.

HEALTH AND SAFETY

Now imagine a less pleasant scenario. A customer gets sick from eating some eggs sold through a farmers' co-op. Is it possible to identify the farm the eggs came from and what happened to them from the time they were laid to the time they were sold? If not, then all of the farms would have to come under scrutiny instead of just one.

If the farm of origin is identified, it should be possible to retrace the eggs' journey to establish if the problem lay with the storage conditions, the chicken feed or water, general living conditions or a sick bird. In each of these instances, if you know what the problem is, you can fix it; if you don't, the whole system may have to shut down.

"When you're growing and marketing something, you need an audit trail. If we sold a bad chicken, we need to know what flock that came from, whether we need to pull the whole flock, whether it was some kind of anomaly or what. If you're selling hay, you need an audit trail. If someone says you sold them mouldy hay, you need to know that it came from a particular field and had been sitting in a pool of water. You need to know why." – Sheila Hamilton, organic farmer, AB

ORGANIC INTEGRITY

A customer purchases dried split peas at a grocery store and wonders if they are really organic as stated on the label. How can they check?

The consumer could contact the certifier whose logo appears on the package. In Canada, customers can also contact CFIA[10] and ask for an investigation. CFIA would contact the certifier to verify the organic status of the peas.

Now the customer wonders whether the product came from that particular farm and whether the farmer didn't simply re-bag cheaper non-organic split peas. Also, how does the certifier know that the farmer did not do anything unacceptable to the product?

These are the types of concerns that the certification and inspection process is designed to address. If the farm has achieved organic certification, the certifier has established organic integrity by doing at least one of the following:

» Yield/ Sales Audit: Do the sales records tally with the harvest records and is the yield appropriate for the acreage planted?
» In/Out Balance: Assuming the farmer buys some organic split peas do the purchased peas plus the harvest totals correspond to the sales plus inventory?
» Audit Trail: Can the peas be traced from the point of sale back to the field in which they were planted, and are all processes that occurred along the way documented?

10 CFIA: Canada Food Inspection Agency, reporting to the federal Minister of Agriculture and Agri-Food

In this case, an organic inspector would check for these records:

» Farm map: indicating the size of the pea-growing area
» Inputs: organic seed documentation, dates and rates of soil amendments, composting records, weed & pest management substances
» Production: Field Activity records of seeding, cultivation, harvesting, equipment cleanouts
» Products: harvest quantities, storage locations, packaging, labels, sales records

To some extent, the system is built on trust. A farmer could conceivably use a banned substance but not record it. If the certifier receives queries or complaints about a farm, a surprise inspection or testing may follow.

TIPS ON IDENTIFICATION NUMBERING

Consistency is essential in the identification of fields, storage facilities, harvest or process lots, and animals; otherwise the complete history will be impossible to trace through your journal, maps, field activity record, control logs, labels and letters.

For livestock, use the animal identification system that you use for other government requirements: livestock tags (with the animal's names if you commonly use them).

For crops, keep it simple. Certifiers may ask that your certification number be part of your lot number format. There are several variables –crop, field number, harvest date, harvest equipment, staff, or weather at harvest, for example –that you could include in the lot number itself, or in a log where lot numbers are recorded.

Record enough information in your lot number log so that it can be cross-referenced to other records for the farm such as:

» Map
» Field/ Bed History
» Input documentation
  - Seeds
  - Soil amendments
  - Pest and disease control

LOT NUMBERING EXAMPLES

Post-harvest handling Log for Cranberry Beans

Date	Inputs	Unit size	Quantity	Lot Number	Supervisor
02/02/08	CB 06	1kg	50	06/07033	Jean Wallace
03/02/08	CB 06	1kg	14	06/07034	Joe Wallace
03/02/08	CB 07	1kg	96	07/07034	Joe Wallace
On the date above,	*Cranberry Beans from harvest year 06 or 07 were packaged*	*in bags of the size above*	*for a total of this number of packages.*	*Each new package had the lot number above, combining the harvest date and packing date*	*on the workday supervised by this person.*

In this example, the log shows the lot number, date, and other details such as package sizes, quantities and personnel involved. The lot number combines the year of harvest (06 or 07) with the Ordinal date of processing[11] (07033 or 07034).

Processing Log for Cranberry Beans Mix

Date	Inputs	Recipe	Unit size	Number of Units	Lot Number	Supervisor
20/12/2007	Cranberries Beans 07 Black Eyed Peas 07 Split Peas (Blue Barn) 1206 Red Lentils (Coopers) 27106	Equal amounts of each by weight	1 kg	600	07354	Richard Lane
6/1/08	Cranberries Beans 07 Black Eyed Peas 07 Split Peas (Blue Barn) 7530 Red Lentils (Coopers) 06271	Equal amounts of each by weight	1 kg	1000	08006	Marissa Ursuliak
On the date above,	*you used your own Cranberry Beans and Black Eyed Peas from harvest year 07 along with purchased ingredients (show their source & lot numbers)*	*in the above proportions*	*to create bags of the size above*	*for a total of this many packages.*	*Each new package had the lot number above, (the Ordinal packing date)*	*on the workday supervised by this person.*

In the Log above, the Lot Number is simply the Ordinal date of processing of each batch. The log provides details of the sources of off-farm ingredients and their lot numbers to allow full traceability.

---

11 An Ordinal date consists of 5 digits: the first 2 correspond to the year; the latter 3 correspond to the day ranging from 001 to 366. 10010 is January 10th, 2010; 11247 is September 4th, 2011. For a full chart for leap and non-leap years, check the website http://www.fs.fed.us/raws/book/julian.shtml

THE INSPECTION PROCESS

"An organic inspector must be the eyes, ears, nose, and guts of the certifier, and ultimately, the consumer."
– IOIA Organic Inspection Manual

An inspection is your opportunity to verify that your farm is planned and operated in accordance with the Standards. The organic inspector uses the onsite inspection to observe your organic plan in action on a particular day. If events occur at other times and cannot be observed, your records must fill in the complete picture.

The organic inspector will also use the inspection to check the traceability of some of your products – tracing their history from source to sale – so your records must prove the organic integrity at all stages of your operation.

RESPONSIBILITIES OF THE FARMER

The inspection can be done at any time during your operational year. Usually it occurs during the growing season. Before your inspection day arrives you must do the following:

- submit a new or updated farm plan, application, and annual fee to the certifier by the application deadline (typically March or April)
- ensure that records are complete and accessible for examination
- note any variations from the farm plan; these can easily happen because of weather conditions, market changes, seed trials, running out of seed in the field or completing the planting with another variety or supplier
- save every scrap of paper that pertains to inputs, activities and outputs and organize them in files or binders
- type, print, or write legibly in your journal and in your records so that both the farmer and the organic inspector can decipher it
- review the Standards as they apply to your operation, and ask the certifier for clarification if necessary. Don't wait until the inspection
- check past communications with the certifier to ensure that noncompliant issues from the last inspection have been resolved and that you can demonstrate and/or provide documentation that you are now compliant
- check certifier newsletters for changes in the PSL or Standards or certifier procedures

RESPONSIBILITIES OF THE ORGANIC INSPECTOR

"The job of the organic inspector is to **verify** *information provided by the applicant and the certifier,* **inspect** *the premises,* **evaluate** *all information and observations,* **inform** *the applicant on organic compliance requirements, according to certifier policies, and* **communicate** *findings to the certifier."*
–IOIA Organic Inspection Manual

- Inspectors verify that the application and farm plan are accurate and complete, and that they match the products and processes that they see on the farm.
- Inspectors inspect all aspects of the operation including crops, livestock, processing areas, storage, equipment, records, inputs/ingredients, borders and buffers.
- Inspectors evaluate evidence concerning your operation's ability to comply with organic standards.
- Inspectors communicate to the certifier their observations, evaluation, and areas of concern. The inspection report summarizes both the strengths and weaknesses of the operation.

As farmers, often our only contact with the certification system is the annual inspection. We have lots to say about the certification system and would like to complain to the organic inspector. We'd like advice on how to improve our crop yields. We'd like to know right away if we are going to be certified this year.

However, organic inspectors have a specific role to play as the eyes and ears of the certifier and cannot give advice. Organic inspectors only report what they observe on the farm. The certifier is the one who makes the decision about certification. If you have a complaint, contact the certifier yourself; the inspector cannot relay your complaint as part of his or her report. Although some inspectors are available as consultants, an inspector would not do an inspection and provide advice to the same operation because of the conflict of interest.

INSPECTION

At some point during your operational season, the certifier or organic inspector will schedule an inspection with you. A typical inspection consists of five parts:

- introduction and review of the organic plan
- farm tour

- records review
- traceability review
- exit interview

INTRODUCTION

During the introductory interview, the organic inspector will review your maps and farm plan as well as any documents that your certifier may have requested as a result of the previous year's inspection. Your maps should show accurate field numbers, locations and crops, which could have changed since the application was submitted.

FARM TOUR

During the farm tour, the organic inspector will evaluate the accuracy of the maps and determine whether what's observed on the farm–the facilities, crops, soil and livestock–is consistent with what's recorded in your plan. Whether your operation is crops, livestock, on-farm processing, or some combination of all three, the organic inspector may be looking for the following:

Farm tour for crops:

- crops and fields
- soil management
- seeds and seedlings
- greenhouses
- weed management
- pest management
- disease management
- water use and tests
- boundaries and buffer zones
- storage for inputs and crops
- inputs: permitted, restricted or regulated, prohibited products
- risks: herbicides, insecticides, fungicides, residues
- equipment
- harvest plans
- post-harvest handling and storage, buildings, equipment
- sanitation and pest control
- packaging and labelling

Farm tour for livestock

- animals
- housing
- feed rations & supplements
- pasture/range
- water
- health management products
- manure management
- milk handling
- handling for slaughter
- animal identification
- storage of inputs and feed
- sanitation and pest control
- equipment
- packaging and labelling

Farm tour for on-farm processing

as for crops and livestock, plus:

- purchased ingredients
- processing facilities & equipment
- sanitation
- pest control
- water use and tests
- packaging & labelling
- storage

RECORDS REVIEW

Your records must provide enough information for the organic inspector to evaluate your compliance with the Standards. They must also demonstrate your ability to document inputs, practices and production history, and therefore provide complete traceability. Here's what an organic inspector looks for:

» records that are clear, accurate, and organized
» records that contain all the information required for your type of operation
» receipts for purchases that are accessible
» logs or journals that are updated regularly
» sales records that are complete and accessible
» lot numbers for field crops and animal identification for livestock that appear consistently on relevant documents
» an overall record keeping system that complies with organic certification requirements

Read the last column of the master chart in Appendix A carefully. It describes what the organic inspector looks for in each type of record.

TRACEABILITY REVIEW

The organic inspector will pick a sales record at random, and work back through your records to trace the entire history of that item. Your records and supporting documents prove your use of acceptable organic management practices, inputs and products, from the time you received the raw material to the time you released the product.

The organic inspector may ask to see your journal, field activity log, various summary activity records and your supporting documentation.

Any deficiencies in the traceability review will be identified in the inspection report; copies of the documents that were examined may be submitted with the report.

EXIT INTERVIEW

The organic inspector will review the day with you to clarify information and share observations. If changes are made to your documents during the interview, you and the inspector will both sign them. Any issues of concern will be reviewed along with the section(s) of the Standards that relate to them.

CHECKLIST FOR INSPECTION DAY

Be prepared to spend several hours with the organic inspector.

» Ensure that you can devote the time and attention needed to complete the inspection.
» Make prior arrangements for someone else to handle work-related tasks and family commitments.
» Have all your records ready and accessible.
» Provide a space where you and the organic inspector can comfortably review records. A tailgate may suffice on a sunny day, but a clear table and place to sit out of the wind and weather are preferable. Some inspectors require space and electricity for a laptop computer.
» Be prepared to provide easy and prompt access to all fields, buildings, and storage areas, both on- and off-farm. This includes having keys to gates and sheds and having other management personnel available. If you have multiple fields or sites, be sure to advise your inspector, so that sufficient time is allotted for your inspection. The inspector will want to see all parts of your operation, whether or not they are to be certified, in order to assess the risks to organic production [12].

AFTER THE INSPECTION

Back at the office, the organic inspector will summarize the inspection and complete any required reports. The inspection checklists and summary will then be submitted to the certifier, and the inspector will file the detailed notes for future reference.

At that point the certifier should have the organic inspector's report, your application and supporting documentation (including updates requested by the inspector), and your certification fee. The Certification Committee can then review the complete file and make its decision on your organic certification status.

This decision could be one of:

- approval
- approval with conditions that must be fulfilled before next year's inspection
- approval with conditions that must be fulfilled prior to certification
- denial (with reasons clearly stated, based on the standards; plus procedures for appealing the decision)
- deferral, needing more information

12 adapted from Baier, Ann. 2005. *Preparing For An Organic Inspection: Steps And Checklists.* ATTRA (www.attra.ncat.org)

Within a single operation it's possible to have one production system certified without conditions, one with conditions, and one denied certification. For instance, your field crops may pass inspection, your pasture may have conditional approval and your livestock denied as a result of the use of growth hormones to put weight on your steers. Or one field may be certified and another not, because of the use of herbicide the previous spring. Certification decisions and their reasons will be clearly stated in a letter. If the decision includes approval, a certificate will be included with the letter.

If you receive a conditional approval, you must take the appropriate actions to correct the non-compliances, and provide documentation to that effect. Major non-compliances may require a second inspection, which must be paid for by the farmer. Many minor non-compliances can be avoided by having all your paperwork in order the first time around.

CERTIFIED

Congratulations, your organic certificate is issued! You can display it when marketing and also provide transaction certificates to buyers who require them. If you have labels for any of your products, the certifier can licence you to use their certification logo. For consumers, the certifier's logo is additional proof of your organic integrity.

You can also apply for a licence from CFIA, through your certifier, to use the Canada Organic logo.

Certificate of Conformity
Canadian Organic Products Regulation

Maureen Bostock & Elizabeth Snyder
1807 Highway 511
Balderson, ON K0G 1A0

Reg. # 02-1026

Pro-Cert Organic Systems Ltd. (Pro-Cert) hereby certifies that the above operation is compliant with the **National Standards of Canada (CAN/CGSB 32.310 and .311)** and the **Canadian Organic Products Regulation 2009 (COR)** as specified in the attached Organic Production Summary which is part of this certificate. This certification is based on the evaluation of: (i) an application containing the organic production plan, (ii) the results of an inspection of the operation, (iii) cropping history, (iv) production records and (v) such tests or other information deemed essential by the certifying agent.

Category of Organic Certification: Greenhouse, Vegetable, Fruit, Herbs

Products Certified Organic:

Greenhouse, Garden Vegetables, Fruit, Herbs, Hay, Pasture

See the attached Organic Production Summary

Effective Period of Certification: August 1, 2010 to July 31, 2011

Initial Date of Certification: December 20, 2002

This certification remains in effect until surrendered by the holder or until the end of the effective period of certification or until revoked by Pro-Cert pursuant to the Organic Products Regulations, 2009. The extension of the certification requires an annual update of the organic production/handling plan, an annual inspection (within the first 6 months following the anniversary date) and annual evaluation by Pro-Cert.

Certification System: #6 according to ISO/IEC Guide 67

J. Wallace Hamm

J. Wallace Hamm, M.Sc., P.Ag.
General Manager

Date of Issue: July 26, 2010

Pro-Cert
ORGANIC

Via the Eastern Branch Office
2311 Elm Tree Road, PO Box 74
Cambray, ON Canada K0M 1E0
Ph: (705) 374-5602 Fax: (705) 374-5604
www.pro-cert.org

CFIA Accreditation # COR00906030

Keep your original certificate on file in a safe place! You should also keep previous applications, as well as certificates and letters from the certifier. These now form part of your record keeping and farm memory and will help you improve your organic practices.

HIGHWOOD CROSSING FARM

TONY AND PENNY MARSHALL, ALDERSYDE, ALBERTA WWW.HIGHWOODCROSSING.COM

FARM:
Field crops and packaged food processing, 320 acres (130 ha)

ORGANIC CERTIFICATION:
QMI

FIRST CERTIFIED:
1993

RECORD KEEPING SYSTEM:
Paper and computer

WEBSITE:
www.highwood-crossing.com

Before there were roads or bridges in pioneer Alberta, Highwood Crossing was the name for a shallow portion of the Highwood River south of Calgary where travellers used to cross. Tony Marshall's great-grandfather established the family farm there in 1899. When Tony and Penny switched back to sustainable organic farming methods in 1989, they were returning to agricultural practices similar to those of the original farm.

Highwood Crossing Farm has been certified organic since 1993. They grow wheat, rye, flax, canola, oats, barley, peas, hay and sweet clover.

The Marshalls have a certified organic processing facility on the farm where they produce and package food products from the crops that they grow. These products include cold pressed flax and canola oil, organic granola, flax seed muffin and pancake mix, stone ground flours, and whole grains and cereals. Browsing through their website reveals lots of nutritional information (Penny is a home economist), delicious recipes, and high praise from chefs across the country who use their products.

Their packaged products are sold to retailers and restaurants across the country, so their record keeping focuses on production records and customer orders as well as field management.

Tony and Penny evolved into roles based on what they enjoy most or have the best skills for. Penny looks after quality control, new product development and on-farm crop rotations. Tony is responsible for management, customer relations, and sales and marketing. Daughters Megan and Kerry helped with production and market sales during their high school years and now return from their current ventures in Calgary when needed. The Marshalls employ a full time production manager, several part-time staff and a neighbour hired for the fieldwork.

As processors as well as producers, the Highwood Crossing office is a busy place. "We originally set the office up on a professional model and gave it the space, equipment and time needed," Penny explains.

Tony, Penny and their production manager update the records weekly for food production and as soon as possible after crop transactions. As orders, faxes and emails arrive, an "Orders for the Week" page on the computer gets updated. The page is printed off each Wednesday, which is their day for production and assembly of orders, and then filed.

"We remember what is going on by the 'current' pile of paper on one area of the desk; once that's processed it is filed appropriately. Of course, things don't always wrap up as quickly as one would hope," says Penny. The 'current' pile is topped by a scribbler for messages and notes. The current pile contains things such as scale tickets from a grain sale that need an invoice created or the paid invoices that need to be filed. A lot of office work involves email. Tony responds to customer inquiries, completes surveys, answers media questions, and refines quotes from suppliers, for example. He stores this information in an email program.

Field work is invoiced seasonally with all the field work activities and dates recorded. Employees keep monthly timesheets. Summaries of payroll records go into a binder for the current year with each employee having their own section. At year end, these are pulled and filed in an archival drawer.

Everything goes into binders: bin records, farm planning, off-farm purchases, batch records, weekly orders, payroll hours, sales, orders that are shipped by Canada Post. Each binder has its own forms. Many are simple

YEAR	FIELD NO. 1	FIELD NO. 2	FIELD NO. 3	FIELD NO. 4	NOTES
2010	Wheat	Hay	Fallow	Clover	
2011	Peas Green Manure	Hay	Oats	Barley	
2012	Rye	Hay	Peas	Peas	
2013	Flax	Hay	Wheat	Wheat	
2014	Fava beans	Hay	Clover	Fallow	
2015					

As the rotation plan for soil-building is implemented, Penny adds information about variety, seeding rate, harvest date, yield and $ per bushel received as well, making the chart a good economic overview of the fields.

grids made on the computer and printed off as blanks for staff to fill in. The production computer is also used for creating labels, invoicing and record keeping.

"Three-ring binders with inside pockets have been essential for us " says Penny. "They're easy to grab and do that one notation. They're also easy for storing info that doesn't go on our forms and simple for others in the office to see and use. We also use dividers inside the binders for organization and clear page protectors for collecting slips of paper."

Anything that is especially for certification also has a binder and files. The farm map is a photocopied satellite map. Field rotations are on paper, with the various crops shown by years.

At inspection time, Tony and Penny do a review of the certification files from the previous years. Then they check bin records and field histories to update the farm plan. "Time spent in filling out application forms clearly and preparing for the organic inspector's visit seems to help the process go smoothly. We take the time to edit our work and make it easy to read. QMI has a variety of created forms that one can personalize instead of creating everything from scratch," says Penny.

She remembers their early paperwork-from-scratch days: "My first mistake was a bin record that we created without an area for IN and OUT and BALANCE. Needless to say, it looked good with harvest dates, yields, producer number, bin size, and so on, except that there was no way of easily seeing what was left in the bin."

Tony's tip for a smoother inspection day is to take photos of field and food operations throughout the year. "It's appreciated by organic inspectors as they often aren't at the farm when actual events are happening," he says. "Unfortunately, we're usually very engaged for those same events so we miss the photo opportunity."

5 GETTING BETTER

> *"You have flexibility to create and maintain records that are appropriate and well-adapted to your operation, but they must disclose all activities and transactions. Make every effort to keep your records up-to-date, well organized, and readily accessible. And remember – good records lead to better management, improved yields, and higher profits."*
>
> *Jim Riddle, organic inspector*

IT IS 8 O'CLOCK ON A DARK WINTER EVENING. Your chores are done and your thoughts turn to the coming season. Sitting by the woodstove with a hot cup of coffee, organic farmers across the country pull out their records from the previous year and start to plan the new season.

- Should I plant oats in the north field this year or should it be put into a cover crop? Where will I get the seed?
- Will we get the wheat combined by Jones this year or purchase our own combine, and leave combine clean-out affidavits behind in the dust?
- What were the successes and failures of our breeding program? Should we change the focus from feet and legs to udder attachment?
- What biological options are available for cucumber beetle control? Last year's strategies weren't satisfactory.
- Can we justify a new tractor next year? The old one is suffering from metal fatigue, and the repair cost is getting too high.

The planning work you do in the winter answers many questions. Some are critical to the viability of the farm business; others are required for organic certification.

Record keeping is not just for the certifier. You can and should use your farm records to evaluate and improve your farm's economic health and the health of its natural resources. Good records help you remember farm activities and natural events that occur over several growing seasons and provide data for analyzing causes and effects.

UPDATE YOUR PLAN

Your organic plan can change the minute after you send in your certification application. Often, these changes present new opportunities or challenges. Here are some typical examples:

- fields added so you can try that new variety of corn
- the use of oilseed radish instead of rye in the rotation because organic seed was not available
- the replacement of poultry after losing most of the flock to foxes
- the need for anthelmintics because a wet summer contributed to higher

parasite numbers in the sheep pasture
- making salsa instead of sun-dried tomatoes because the dehydrator broke

Other changes may be deliberate choices to add or discontinue a particular crop or breed, pursue a new market, or try value-added processing with your farm products. Many changes will require planning to ensure that the new operation complies with organic regulations and that no new risks are created.

One of the main reasons for having an organic plan is that you'll have strategies for dealing with situations such as those listed above when they occur.

These strategies should then be recorded in your Farm Journal and in other appropriate documents. From there they can be fed back into your next organic plan ready for the inspection and for the upcoming season.

Some changes occur between the time you submit your application and the time of your inspection. In that case, have an amended version of your application/organic plan ready for the organic inspector, as well as an updated Field History or other supporting documentation.

The inspection may highlight some areas for improvement in your methods. Read the Standards again, seek clarification from your certifier, and talk with experienced growers. Keep notes on what you do, both in your Farm Journal and on the appropriate forms, and update next year's application/organic plan accordingly.

Tweak your record keeping

"We're constantly creating new and more user-friendly formats – better for the certifier and better for us."
Rod Randen, organic farmer, AB

Many of the farmers interviewed for this book emphasize that record keeping is an evolving art. Start with the templates and examples presented here or with the forms provided by your certifier. Adapt and refine them for your own operation as you go.

Here are some other ideas:

- Start simple. Pick a minimum number of templates to begin with, and get used to keeping regular notes in your journal and activity logs. After a season or two, you will know whether you are keeping enough information for your own needs as well as for certification.
- Be consistent with identification numbers (or names) for fields, storage bins, animals, etc; otherwise you'll have trouble comparing from one year to the next.

Most of the records we keep are in the farm file, not the inspection file, because we don't just keep them for inspection purposes. When we designed our records, it wasn't for the inspection, but more for what information we need to keep as a business. Whether or not we're certified, we'd still keep the same records."
Rowena Hopkins, organic farmer and inspector, NS

- Keep thorough notes, using paper or computer, on each crop, the varieties planted each year and their performance, the source of seed, and anything done for fertility or pest and disease management. If you grow many varieties of tomatoes, for example, it's easy to lose track of which one matured earliest *and* tasted the best. The crop source, organic status and crop/field management data will all be needed for inspection, but this information will also help with your decision-making for next year's crop planning.

Advice from organic inspectors and certifiers

In this section, we have collected bits of wisdom about inspections and about record keeping from some experienced organic inspectors.

Janine's top ten tips for a great inspection

Janine Gibson, IOIA organic inspector and trainer.

Application forms

No. 10 Print clearly and answer ALL the questions...use "n/a" rather than leave blanks...but think outside their box...how might it apply to your operation?

No. 9 Put your best foot forward. Brag about what you are proud of doing on your farm.

No. 8 Call your certifier and ask if you are unsure of what they mean by a question. Every question is related to something in the Standards. Understand what it is.

Paperwork

No. 7 Be ready with your production records; have them laid out on the table for review. We all want the info you provide to be accurate.

No. 6 Calculate your total acres of each crop. There are almost always changes from the time you submit your application to the time of the inspection. Have those changes listed and ready to be added to your organic plan.

No. 5 Have all your sales records back from the accountant.

Tour

No. 4 Know your tour route hazards–walking and driving.

No. 3 Have spare boots and/or boot covers ready, especially in mucky weather.

No. 2 Have enough gas in a working vehicle that is appropriate for the terrain.

No. 1 Provide clear, specific, up to date directions to your farm or facility. A lost inspector is a cranky inspector.

JIM'S 9 TIPS TO SHORTEN YOUR INSPECTION TIME

Adapted from Jim Riddle, 2004. The Inspector's Notebook no. 1, Rodale. http://newfarm.rodaleinstitute.org/columns/inspector/2004/0804/9_tips.shtml

For me, interviewing the farmer is the best part of being an organic inspector. I get to hear the stories of organic farmers firsthand – their successes, their failures, and the pride and excitement they feel in what they are doing.

I try to make the inspection a good experience. I want the farmer to get something out of it. At the same time, I have to make sure the information is accurate and the paperwork, complete. Otherwise, certification can be delayed or even denied.

One memorable inspection finished at midnight and was no fun for me or the farmer. We had started the inspection at 6:00 pm. After inspecting numerous fields and locating the beef herd in a distant pasture, we sat at the kitchen table to complete the paperwork.

The reason it took so long was that the 50+ fields had had acreage and number changes (and new maps) since the farmer had completed his organic plan. There were also some changes in the crops he had planted compared to what he had planned to plant – a common occurrence. He had not prepared a new acreage list or updated the farm map. As a result, we spent about two hours redoing his "crops requested for certification" table. Accurate information is required for organic certification.

Many inspections take longer than necessary, often because the farmer is not prepared. Other reasons include: previous certification conditions have not been addressed; maps and field histories are not up to date; records are not accessible; seed and other product labels were not saved or are not accessible; organic harvest and sales records are not available; or records are not organized to track products from sale back to the field(s) of production.

Other factors can delay the inspection. For example, farmers often keep old products or containers from prohibited substances that were formerly used on the farm. When these containers are around it raises additional questions and extends the length of the inspection.

HERE ARE NINE SIMPLE IDEAS TO SHORTEN THE INSPECTION TIME:

1. **Read the letters you get from the certifier.** During the inspection, I review all previous conditions for certification and issues of concern identified by the certifier. These letters clarify what issues I will be discussing with you, so be prepared to respond.
2. **Organize your records** in file folders or a ring binder with separate sections. Have your records readily accessible for the inspection.
3. **Provide accurate and up-to-date written information.** If your crop plans, planting dates, field sizes, crop locations, or input plans have changed since the organic plan was submitted, be prepared to provide written documentation to the inspector.
4. **Verify your acreage.** The total of individual organic field acreages must equal your total acreage requested for certification. To be sure, add up your individual field sizes and compare it to the total acreage requested for certification. Do this *before* the inspection.
5. **Keep all labels and receipts.** This includes seed bags or packets, soil mix ingredients,

and all inputs including fertilizers, pest control products, animal health care products, animal feeds, and feed supplements.

6 **If you use a lot of inputs, use an input inventory sheet** (with columns for the brand name, source, location used, date first used, date when approved by the certifier, and date discontinued use). This helps clarify which inputs are used and whether they have been approved. If an input has already been approved, the inspector does not have to write down all label information and consume valuable inspection time. Input inventory sheets can be used for fertilizers, pest control products, and animal health care products.

7 **If you sell bulk commodities, keep accurate harvest totals and sales records per crop.** The organic inspector will choose one of your crops to do a "sample audit balance" to see if your records can track the crop from sale back to the field(s) where it was grown and to assess if the amount you harvested and sold was realistic for that crop.

8 **If you are a market gardener or if you sell products directly to consumers, keep daily or weekly sales totals**, along with accurate input and production records. The inspector will need to review your production and sales records.

9 **Walk through your storage areas** before the inspector arrives and get rid of any products you no longer use. Check with your local Extension Service or municipal office for proper disposal options.

Thanks for being prepared. It helps me do a better inspection and facilitates your organic certification.

DEBBIE'S POST-INSPECTION TIPS

Adapted from Debbie Miller, OCIA. Ask a Certifier. The Canadian Organic Grower. COG, Fall 2008.

"My organic inspection is over and it went well. Now, what can I do to make my certificate arrive faster?"

First, take another look at the affidavit your organic inspector left behind. Did he/she note any areas of concern? Was there any missing paperwork? If so, take care of this right away. Ideally, you want to send the information to your certifier even before they receive the report noting that it is needed. If you get a fax, letter, email or phone call from your certifier requesting more information or clarification, respond immediately.

Missing information is the most common reason certificates are delayed, and it is usually very simple information that we need – an updated map or an input label, for example. Certification reviewers can't make a decision until they have all the information.

Double-check that your certification payments are up-to-date. By the time the inspection is finished your certifier has already invested a lot of time and money on your file, and has paid the inspector on your behalf, trusting that you will meet your financial obligations. If you are behind in your payments it is likely that your file will be held, waiting for payment.

Normally, files are reviewed in the order in which they are received: first in, first out. If you have sent in everything that was requested of you, your payments are up-to-date, and you have a sale pending receipt of your certificate, you may call your certifier to request that your file be moved closer to the top of the review pile. Please don't do this unless it is truly an unusual and urgent situation; if everyone requests that they be moved to the top, we'll be back to where we started.

Resources for planning and record keeping

The following list of resources is in no way definitive. Books, articles, computer programs and websites are being developed all the time; only a few are mentioned here.

BOOKS

These planning books include advice on record keeping:

Crop Planning for Organic Vegetable Growers. Frédéric Thériault & Daniel Brisebois. Canadian Organic Growers, 2010. The planning method and templates you need to ensure that a variety of crops are ready each week for your markets and home delivery customers. Includes seed, harvest and sales records that show traceability.

Building a Sustainable Business: A Guide to Developing A Business Plan for Farms and Rural Businesses. Minnesota Institute for Sustainable Agriculture, 2003. Follows a new dairy operation from their initial exploration of values, brainstorming of goals, and research into detailed planning for on-farm milk processing, markets and financing. Available online at http://www.misa.umn.edu/vd/bizplan.html or in print format from many book sellers.

The Organic Farmer's Business Handbook. Richard Wiswall. Chelsea Green, 2009. Based on 27 years of experience on a mixed vegetable operation, advice to make vegetable production more efficient, better manage employees and finances, and turn a profit.

SOFTWARE

"Farmers don't want to write everything down twice. Computer records should be a condensed or organized form of the farm journal, not a duplicate. We don't have time in the growing season to be entering things on the computer; our paper forms need to capture enough information so that we can transfer them later, in summary form."

Rowena Hopkins, organic farmer & inspector, NS

Paper forms are easy to carry around for instant updates, but computerizing at least some of your records will be a major time-saver when you're trying to locate or analyze information, and when generating forms with information that stays the same every year. There are many programs and formats available; both general-purpose and purpose-made software are fine.

David Methot in New Brunswick recommends free software for spreadsheet and note-taking applications, 'Open Office' www.openoffice.org and 'SEO Notes' www.SEOnote.info respectively. He uses note-taking software for varietal information and has built up a hierarchy of information which starts with a list of the vegetables and fruits that the farm grows. Each plant-type expands into a list of cultivars tried, growing strategies used, etc. The spreadsheets in 'Open Office' are used for more quantitative records such as amounts grown, harvested and sold, expenses and income. Other free software and for-purchase software is available as well; the most commonly used for-purchase spreadsheet software is Microsoft Excel.

For crop farmers, consider Field Manager PRO from Farm Credit Canada (FCC) – see Mark Bernard's description in the Barnyard Organics farm profile. It keeps field histories, compares crop scenarios, tracks information for government programs, and links to FCC's AgExpert Analyst for financial management. www.fccsoftware.ca

ONLINE

Many record keeping forms, templates, advisory bulletins, resources and links are available online from various sources. Here are a few:

Canadian Organic Growers (COG) www.cog.ca – computer formats for the forms contained in this book as well as *Crop Planning for Organic Vegetable Growers*

Canadian Food Inspection Agency (CFIA) www.inspection.gc.ca – follow the links for Food/Organic – then find information on the Regulations, Standards, Certifiers, labelling requirements, imports and exports, and the Standards Interpretation Committee. Follow the links to the Certifiers websites to obtain access to their record keeping templates.

Atlantic Canadian Organic Regional Network (ACORN) www.acornorganic.org – Check especially "The Organic Path" for steps and resources in transitioning to organic growing. ACORN also has a brand names directory.

Certified Organic Associations of British Columbia (COABC) http://certifiedorganic.bc.ca Lots of regional and generally applicable resources here. Check under "Cyberhelp" for "certification" resources.

National Sustainable Agriculture Information Service (ATTRA) www.attra.org – Many downloadable bulletins and forms for records and growing information for specific types of operations.

Organic Trade Association (OTA) www.howtogoorganic.com – follow the "Pathway for Producers" for many links on transitioning and certifying

Rodale / New Farm www.rodaleinstitute.org/new_farm Lots of resources including the Rodale Institute's electronic Organic System Plan tool, which helps you assemble the necessary documentation to apply for organic certification through a certifier. Written for US/NOP Standards, but generally applicable to other countries. Designed to be used in conjunction with The Rodale Institute's Transitioning to Organic online course, but also can also be used independently.

University agriculture extension departments, Government agriculture departments: Check with your nearest government or university agriculture department; many have bulletins, forms and sample plans available on their websites.

"The biggest change is between your ears."
Pieter Biemond, organic farmer, ON

Sunworks farm

Ron and Sheila Hamilton, Armena, Alberta www.sunworksfarm.com

Farm:
Livestock,
240 acres (97 ha)

Organic certification:
QMI

First certified:
1997

Record keeping system:
Paper and computer

When your farm sends 2000 broiler chickens per week to market, you need a pretty good tracking system. Sunworks Farm is the largest certified organic poultry producer in Alberta. Besides the meat chickens, Ron and Sheila Hamilton care for 1500 turkeys, 2500 ducks, over 1000 laying hens (more in summer), as well as 50 cows and heifers.

"We bought the farm in 1992, started conversion in 1993," says Ron. "Our first certified organic year was 1997, with 100 broilers and two cows. The operation has grown 30% each year for the last 5 years." Sunworks sells to three restaurants, two year-round farmers' markets in Edmonton and Calgary, at the farmgate and at their son-in-law's store in Calgary.

"I used to do my flow chart for the chickens on a calendar with flock numbers and age each week," says Sheila. "Now to avoid the manual rewrites, I have a computer spreadsheet with my flow chart on it, colour coded. I can go into that and indicate which birds are in brooders or field shelters. Every three to four weeks I print a new flow chart because we've changed where birds are going or where flocks are coming in."

Sunworks employs seven full-time and one part-time staff. One of the challenges of a large operation is quality control. Sheila stresses that staff must be trained right from the beginning about the importance of traceability and record keeping so that there are no gaps. They encourage their employees to develop their own records as well. Sheila has noticed that beginners tend to make it more complicated than it needs to be.

The farm records have evolved from notebooks, calendars and binders to the computer. Sheila and Ron have created their own spreadsheets to include the information that the certifier needs and organized in the way they need it for the farm. "It's basically columns and lists," says Sheila. "It can be simple and still have all the information you need."

Paper forms are filled out daily in the barns, field shelters and feed mill, and then summarized on the computer. The livestock records include calving, CCIA tags, care and feeding, and pasture rotation, all of which are set up on spreadsheets. There are flock sheets in every barn where staff keep a daily journal for each flock: temperatures inside the barn and out, when shavings are put down, when the flock was moved to a field shelter, how many pails of feed, water consumption, cleaning, losses, etc. There are also cleaning records for the barns, shelters and equipment. They are also involved with the Alberta Chicken Producers On Farm Food Safety program and the Alberta Egg Producers Start Clean Stay Clean program.

The feed mill is CFIA-registered, so the information recorded there includes purchases, production mixes and sales, and is used for CFIA audit trails as well as for organic certification.

Ron obtained aerial photos free from their county office to use for the overall farm map and field maps. They hand-draw the boundaries and colour-code the fields. Their Field History is on the map, showing 5-6 years of where turkeys, broilers and laying hens were throughout each year, all colour-coded. The birds are moved each year from south to north and the shelters are lined up east-west.

Test results for water and manure management are filed for inspection. The office staff use Simply Accounting software to record grain and feed purchases and sales as well as market sales.

As the farm grew and their paper records overflowed from four or five binders, their management system evolved into computer records, with paper backups in two file drawers. At inspection time, Sheila has all the necessary documents printed and ready for the organic inspector to take.

Sheila and Ron are passionate about their commitment to organic production and their responsibility to their customers. Sheila says she's heard other vendors at market tell customers, 'I'm organic, but I'm not certified because it's too expensive, and I don't have time to do the paperwork.' "Customers then come over to me, and I say, 'Well, I do the paperwork.' How dedicated are you? How important is it to you? If you're dedicated, you'll do the paperwork."

"Everything I'm doing right now, I'd be doing anyway. There's no extra record keeping for certification."

Appendix

A Records master chart

The organic plan (chapter 3) introduced four categories for planning and record keeping. This appendix describes the records and plans that are required in each category:

- Physical Layout... page 95
- Inputs... page 96
- Activities... page 99
- Outputs... page 102

This master chart is meant to be used as a checklist when setting up record keeping or preparing for inspection.

- The first column lists subjects that require records.
- The "Records" column lists the types of ongoing records or supporting documents that must be kept for each subject. These records require regular updates throughout the year, as described in chapter 2. Records marked with an asterisk * have templates supplied in Appendix B.
- The "Application / organic plan" column lists the sections of the organic plan or Certification Application that correlate to each subject. This information will be completed or updated once a year, as described in chapter 3.
- The "Inspection" column gives some hints as to what an organic inspector might be looking for in the plan or records.

Physical layout

Subjects	Records	Application / organic plan	Inspection
Farm Maps		Farm Location	Map gets the inspector to the farm easily!
		Farm Layout	Maps are accurate and consistent with field shape, size, location, crops, adjacent land use, buffers, natural features (water, hills, rock). Maps are updated as needed.
		On-farm Processing flow chart and/or floor plan	Processing flow avoids risk points for contamination.
	Neighbouring Land Use		Affidavits, letters, or journal notes explain any crops and treatments that are potential contaminants (e.g. pollen from GMO crops, spray draft)
	Prior Land Use		If land is certified less than 3 years, affidavit shows prior crops and soil treatments.
Field history/ activity record	Field Activity*	Field History*	Crops observed in field correspond with field history/activity record for current year. Record shows that the organic plan was carried out with respect to crop rotation, planting, inputs, field management. Exceptions are explained.
Crop Rotation	Journal	Rotation Plan	Rotation Plan shows effective soil building and crop diversity. Exceptions to plan are recorded in the Journal with reasons.

INPUTS: CROPS

SUBJECTS	RECORDS	APPLICATION / ORGANIC PLAN	INSPECTION
Seeds, seedlings, tubers etc	Seed labels; receipts or delivery slips for seed and inoculants; organic/GE status; seed search log; harvest & storage records for seed produced on-farm	Seed Search, list of seed to be purchased (including cover crops, pasture grass, forages)	Seed is organic. If a non-organic version is approved by the certifier prior to seeding, it must be non-GE and untreated.
Fertility Inputs, Soil	Product labels Soil test results, Organic/GE status	Fertility management plan	Inputs are listed in the PSL and used as per any restrictions mentioned there. Documented plan for crop rotation for fertility.
Compost	Composting log* or receipts and documentation for purchased compost	Compost inputs and/or sources	Sources are compliant with PSL.
Manure	Receipts and source documentation	Sources	Manure sources are compliant with the PSL.
Organic pesticides and herbicides	Product labels, application history (in journal or field history)	Pest and weed control strategies	Inputs do not contain prohibited ingredients. Restricted ingredients are used only when necessary and according to the PSL.
Mulch	Products labels, delivery slips or letter from the supplier	Sources and management plan	Plant based mulch is organic. Plastic mulch is biodegradable, recycled or re-used. No GE content in biodegradable mulch.
Pest control products	Application notes in journal or Field History	Pest management plan	Products and applications are in accord with the PSL.
Water	Water tests (where necessary)	Water sources	Crops are not at risk from contaminated water sources (pathogens, heavy metals etc).

Inputs: livestock

Subjects	Records	Application / organic plan	Inspection
Purchased Livestock	Receipts of purchase, Organic documentation for livestock	Plan for livestock replacement or additions.	Purchased livestock is organic or necessary for genetic diversity when building a herd.
Feed	Product labels, receipts, delivery slips and documentation of organic status	Sources	Feed is certified organic.
Feed Supplements	Product labels or receipts; MSDS, ingredient list	Nutrition management plan (Feed rations), Sources	Supplements do not contain prohibited ingredients. Restricted products are used only when necessary and according to the PSL.
Health Care Products	Herd / Flock Health Log*, Product labels, receipts	Sources, anticipated problems and responses	Products do not contain prohibited ingredients. Restricted products are used only when necessary and according to the PSL.
Bedding	Receipts and documentation of organic status	Sources	Bedding that is eaten by livestock is organic. All bedding is free from prohibited substances.
Pest Control Products	Labels, receipts, MSDS, journal/ history of application	Sources and plans for use	Products and applications are in accord with the PSL.
Cleaning Products	labels, receipts, MSDS/ ingredients	Sources, description of normal use	Products and applications are in accord with the PSL.

Inputs: on-farm processing

Subjects	Records	Application / organic plan	Inspection
Ingredients	Packets, receipts of purchases, organic transaction certificates or delivery slips	Recipes	Ingredients are organic. The same ingredient is not being used in an organic and non-organic form. Non-organic ingredients do not exceed the allowable limits.
Food additives and processing aids	Product labels or receipts of purchase	Recipe	Prohibited ingredients are not used. Restricted ingredients are used only when necessary and according to the PSL.
Pest Control Products	Journal, or pest management log*	Pest control plan	Prohibited products are not used. Restricted products are used only when necessary and according to the PSL.
Cleaning Products	Product Labels including ingredients	Sources	Prohibited products are not used. Products are used according to the PSL.
Packaging	Product labels; Food Grade Status; Non-GE statements for biodegradable products	Sources and rationale	Packaging is food grade, is not treated with prohibited products (fungicides, etc) does not contain GE materials and is not used in excess.

ACTIVITIES: CROPS

SUBJECTS	RECORDS	APPLICATION / ORGANIC PLAN	INSPECTION
Crop Rotation	Field History/Activity* records.	Rotation plan, Field Maps	Crop rotation is sufficient to reduce pest, disease and weed pressure and to build soil fertility. Maps are up-to-date.
Buffers and contamination	Previous land use affidavit; Letters to/from neighbouring conventional farms etc	Map, including buffer zones indicating neighbouring land use	Crops are not at risk from neighbouring or previous land use. Affidavits are signed.
Pests	Journal or pest control log*	Pest management plan	Pest pressure is managed effectively without restricted or prohibited inputs.
Diseases	Journal or disease control log*	Disease management plan	Disease pressure is managed effectively without restricted or prohibited inputs.
Weeds	Journal or weed control log*	Weed management plan	Weed pressure is managed effectively without restricted or prohibited inputs.
Manure	Rates and dates of application (journal)	Manure management plan	Manure is not applied within 90/120 days of harvest, or to wet or frozen ground or at excessively high rates.
Compost	Composting log*	Composting process and recipe	Process allows for breakdown of pathogens and restricted content.
Water	Journal; Irrigation log*	Irrigation systems and water management plan	Water usage is sustainable; no erosion or over-use is observed.
Equipment	Clean out Log* or organic status documentation from owners/ borrowers of shared equipment	Equipment management plan, Equipment ownership	Land and crops are not at risk from contamination or commingling with non-organic crops, soil etc
Harvest	Journal	Harvest system	Crops are not at risk from contamination or commingling with non-organic crops.

Activities: livestock

Livestock operations must include all elements of Crop activities (above) as well as Livestock activities listed in this section.

Subjects	Records	Application / organic plan	Inspection
Livestock Numbers	Herd/ Flock numbers/ Inventory	Stocking densities	Density of livestock complies with the Standards. Production levels correspond to livestock numbers.
Livestock Breeding	Breeding records connecting offspring to both parents	Breeding plan	Number of animals corresponds to breeding records. Non-organic breeding stock is easily identified. Genetic diversity is maintained. No cloning or genetically engineered livestock.
Accommodation / Living Conditions	Changes to maps or grazing plan; outdoor access log	Map including access to water, food and shelter; Rotational grazing plan	Livestock living conditions are compliant with the Standards. Livestock have adequate access to the outdoors or pasture. There is sufficient pasture for herbivores.
Physical Alterations	Journal/Log or Alterations Record	Description of processes and inputs	Processes and inputs used are compliant with the Standards.
Health Care Methods	Journal/Log, Herd/ flock health log* or individual animal health records; monthly bacteria counts and SCC reports	Description of processes and inputs	Processes and inputs used are compliant with the Standards.
Parasites	Treatment log, observations	Management plan	Plan in place to minimize parasites; includes pasture management and faecal monitoring. Use of parasiticides complies with Standards for allowed substances, number of treatments, withdrawal periods, etc.
Slaughter	Slaughter records	Abattoir & transport plans	Slaughter records correspond to herd/flock size. In the case of non-organic facilities, systems prevent commingling with non-organic products and contamination with prohibited substances.
Milk/Egg Handling	Milking/ egg gathering records; Egg inventory*	Description of processes	Milk yields/ egg production corresponds to herd/ flock size.
Dead-stock Management	Mortality Records	Dead-stock management processes	Livestock mortality is not excessive. Dead-stock are handled according to CFIA regulations.

ACTIVITIES: ON-FARM PROCESSING

SUBJECTS	RECORDS	APPLICATION / ORGANIC PLAN	INSPECTION
Equipment and facilities	Changes to the map	Map of facilities including location of processing equipment	Systems in place prevent commingling with non-organic ingredients and contamination with prohibited products and human pathogens.
Processing	Processing log	Flow chart of process	Systems in place prevent commingling with non-organic ingredients and contamination with prohibited products and human pathogens.
Packaging	Packaging log	Packaging systems	Systems in place prevent commingling with non-organic ingredients and contamination with prohibited products and human pathogens.
Cleaning	Cleaning Log	Cleaning schedules and processes	Cleaning systems prevent contamination with human pathogens and with cleaning product residues.

Outputs: crops

Subjects	Records	Application / organic plan	Inspection
Harvest Yields	Harvest Log*	Estimated yields	Harvest can be traced back to field(s) of production. Yield is consistent with plan and area planted.

Outputs: livestock

Subjects	Records	Application / organic plan	Inspection
Meat Production	Slaughter and Butchering Records	Expected production levels	Slaughter and Butchering records correspond to herd /flock size.
Egg and Milk Production	Milking / egg gathering records; Egg inventory*	Expected production	Milk yields / egg production corresponds to herd/ flock size.

OUTPUTS:

ALL OPERATIONS – CROPS, LIVESTOCK, ON-FARM PROCESSING

SUBJECTS	RECORDS	APPLICATION / ORGANIC PLAN	INSPECTION
Inventory	Inventory records		Inventory records correspond to harvest records and sales records. (Inventory can include seed saved for next year.) Sales or inventory records cover ALL crops or animal products: organic, transitional and conventional (including buffer-zone crops).
Storage	Clean-out log	Storage facilities map and systems	(Bins for crops; freezers for meat; cooling equipment for milk and eggs, etc) Storage areas are managed so as to prevent risk of contamination or commingling.
Sanitation and pest control	Journal, cleaning log, pest management log	procedures	Sanitation and pest control (including rodent control) products are on permitted list and properly used to avoid contamination of product.
Transportation	Transport truck clean out affidavit*	Ownership, management and cleaning procedures for transportation equipment	Transportation equipment is managed so as to prevent risk of contamination or commingling.
Sales	Sales records; Receipts; Market records*; Bills of lading; Transaction Certificates	Customers, Marketing systems, Anticipated sales	Sales correspond to Harvest or production records; exceptions are explained. Sales or inventory records cover ALL crops or animal products: organic, transitional and conventional (including buffer-zone crops).
Marketing	Photographs of products; copies of marketing materials; web address	Advertising and marketing materials	Marketing claims are true and compliant with the Standards.
Labelling	Sample labels	Label design	Labels comply with the Standards.
Lot numbers	Lot numbers used in record keeping system	Lot numbering system	Allows for traceability from point of sale back to point of production.

Appendix

B Records templates

- Field History / Field Activity
- Composting Log
- Herd / Flock Health Log
- Pest Management Log (processing)
- Pest Control Log (crops)
- Disease Control Log (crops)
- Weed Control Log
- Irrigation Log
- Equipment Clean out Log
- Transport Clean out Affidavit
- Egg inventory
- Harvest log
- Market records

GENERAL INSTRUCTIONS:

Replace header words in italics with your own information. Make enough copies of blank sheets for the year, and be sure to keep the original for making more copies. These forms are available for download from the COG website www.cog.ca under Publications and Practical Skills Handbooks.

FIELD HISTORY / ACTIVITY:

This is offered in two formats: by field or by year. If you have five fields or less, use the "by year" format. Replace *Field Identifer and area* with your own field names and sizes; e.g. "North Field 40 acres."

FIELD HISTORY BY YEAR:

This format has fields across the top, years down the side. You can use one page per year, or continue one year after another on the same page by repeating all the rows. When you are using the spreadsheet templates (available on the COG website) for the Field History, you can insert new rows just under the headings so that the current year is always at the top, and the previous years are below, oldest at the bottom.

FIELD HISTORY BY FIELD:

This format has years across the top, fields down the side. You can use one page per field, or continue describing several fields on the same page by repeating all the rows.

COMPOST LOG:

When each pile is "finished" use a row to record when and where it was used.

PEST MANAGEMENT LOG:

Your regular monitoring (as described in your organic plan) should be recorded, not just actual pest sightings and treatments. Use the last column "Applied or monitored by" to record the name of the individual responsible, or the pest-control contractor.

EQUIPMENT CLEAN OUT LOG:

Substitute the name of your piece of equipment for the *Equipment* in the header. Keep a laminated copy of the Clean out instructions with the Log.

EGG PRODUCTION & SALES:

Make a copy for each month. Substitute the month name for *month* in the first column header. Replace *units* in the column headers with the units that you use: singles, dozens, hundreds...

HARVEST LOG:

This can also serve as a storage/bin record, and as an audit control summary of all production and sales.

MARKET RECORDS:

Market gardens can use the Harvest Log above for harvest and sales, or this version which is more suitable to multi-crop operations. The "end" column is used for the quantity left at the end of a market which is useful for planning how much to plant and harvest. This form is reproduced from "Crop Planning for Organic Vegetable Growers," Canadian Organic Growers, 2010

Field history & activity by year

Year	Field identifier and area	Field identifier and area	Field identifier and area	Field identifier and area	Field identifier and area
	Crop	Crop	Crop	Crop	Crop
Fertilizers and Amendments					
Pest Management					
Disease Management					
Weed Management					
Field Work					

Field history & activity by field

Field and area	Year-4	Year-3	Year-2	Year-1	Current year
	Crop	Crop	Crop	Crop	Crop
Fertilizers and Amendments					
Pest Management					
Disease Management					
Weed Management					
Field Work					

Compost log

Date	Ingredients added	Pile turned?	Temp (°C)	Notes

Herd/flock health log

Date	Observations	Livestock ID	Treatment method and amounts	Products used

Pest management log

Date	Observations	Livestock ID	Treatment method and amounts	Products used	Applied or monitored by

Clean out log for *equipment*

Date	Crop & lot no. or animal ID	Inspected prior to use	Cleaning procedure followed (or describe alternate)	Products used	Signature

---

Transport clean out affidavit

Farm name:

Date transport unit loaded: ____________________

Arranged by: ❑ Grower ❑ Buyer ❑ Other:

Type of transport: ❑ Farm wagon ❑ Farm truck ❑ Bulk semi-trailer ❑ Common carrier ❑ Tanker ❑ Other:

Is this transport used only for organic products? ❑ yes ❑ no

if no, state prior product carried:

Unit found free of: ❑ Odor ❑ Residue ❑ Conventional product ❑ Any substance compromising organic integrity

Transport unit ID	Crop & lot no. or animal ID	Cleaning method (check all that apply, or describe)					Products used
		Swept	Vacuum	Air blown	Washed	Other	

Signature: ______________________________

Egg production & sales

Month	Collected (units)	Sales (units)	To	Notes
1				
2				
3 … 31				

Harvest log

Harvest					
Date	Crop	Field	Quantity	Storage bin/lot no.	
Sale					
Date	Purchase order	Bill of lading	Buyer	Quantity	Value

Market records

Market				Total $ sold			Date
Crop	Unit	Price	Planned harvest	Actual harvest	End	Sold	Notes

Appendix

C Excerpt from *A Guide to Understanding the Canadian Organic Standards*

Prepared by Canadian Organic Growers, January 2010

The complete Guidance Document (and updates) is available for free by request from COG (see inside back cover for contact information). This is not a legal document. Our interpretation can help you to understand the intent of the COS, but it is not authoritative.

The following adaptation is the guidance on Paragraph 4 of the Canadian Organic Standards, organic plan, Record Keeping and Identification:

In practice, most certifiers supply operators with an application form that, once completed, becomes the 'organic plan' for the operation. See also Paragraph 6.7.9, which requires a parasite management plan for livestock operations and Paragraph 7.1.6, which requires an organic plan for apiculture operations.

Generally, an **organic plan for crop production** includes:

1. a detailed map of the operation, a description of the rotation plan and production plan, a description of changes in the general condition of the soil and ongoing monitoring of the soil condition, field histories documenting previous crops, inputs used, and transition status;
2. a detailed description of the sources of seed, including seed inoculants, germ plasm, scions, rootstock and other propagules;
3. a description of the cultivation techniques and types of machinery and equipment used; a profile of erosion risks and proposed corrective measures;
4. a description of the fertilization program, including origin and source of manure, storage and handling techniques, quantity applied, application period and composting methods; a description of other production methods aimed at increasing organic matter, such as green manure crops and harvest residue management, and a plan to prevent the leaching of breakdown products of liquid and solid manure;
5. a detailed listing of all production inputs (including pest control products) and the justification for their use;
6. a description of the watershed on the operation and the measures to prevent its exposure to or contact with prohibited substances; a description of the sources and quality of water used for irrigation;
7. a description of crop protection issues and management strategies; a description of problems with past practices, if applicable;
8. a description of potential sources of exposure to or contact with prohibited substances; concerns associated with neighbouring areas and buffer zones; in cases where the operation is not fully converted to organic pro-

duction, a description of the management system to maintain organic integrity;

9. a description of the facility's management plan for the storage and handling of organic inventory, and the steps or procedures taken to prevent the commingling of organic and any non-organic stocks that may be present; and
10. for wild crops, a detailed plan of the harvest areas of wild plant species and a history of the last 36 months of compliance with these standards, including a description of the harvesting methods used and the proposed measures for protecting wild plant species.

Generally, an **ORGANIC HANDLING AND PROCESSING PLAN** includes:

1. a description of all specifications and steps under the control of the operation, such as the harvesting, preparation, packaging, labelling, processing, storage and distribution of organic product;, as well as a description of the controls required at other levels of processing and/or handling (in the supply chain) in order to maintain the status of organic in accordance with the requirements of these standards;
2. a schematic flowchart or written description with sufficient information for a general understanding of the flow of organic products during handling and processing;
3. a description of a control system for processing and/or handling that addresses the prevention of commingling of an organic product with non-organic products and potential exposure or contact through the following:
   i. ingredients;
   ii. containers and packaging;
   iii. enzymes;
   iv. pest control substances;
   v. prohibited handling and processing procedures, such as the use of food irridation;
   vi. sanitizers, boiler chemicals, lubricants, processing aids and prohibited substances; and
   vii. transportation and storage;
4. listing of all material inputs including:
   i. all ingredients and substances used in handling and processing of organic and non-organic products;
   ii. for each product labelled organic containing one or more non-organic agricultural products as ingredients, a written description of the rationale for not using organic ingredients; the efforts made to locate or develop a source of the form of the organic ingredient, in accordance with these standards; and progress made over the previous year(s) to eliminate non-organic agricultural products as ingredients;
   iii. processing aids used; and
   iv. description of water usage in the handling operation.
5. listing of all pest management inputs including:
   i. pest problems encountered in the handling operation, and pest monitoring techniques employed;
   ii. pest management plan;
   iii. nonchemical pest control methods used in the handling operation; and
   iv. chemical pest control methods used in the handling operation.

6. description of livestock handling practices, where applicable (abattoirs), including:
   i. a health plan identifying nutrient supplementation;
   ii. handling methods used to minimize stress in livestock;
   iii. arrangements made at the packing plant for supplying livestock with fresh water; and
   iv. arrangements made at the packing plant for feeding and bedding livestock held for more than 24 hours;
7. description of waste management practices including:
   i. efforts taken to reduce solid and/or liquid waste and airborne emissions produced by the handling operation; and
   ii. recycling efforts, such as the use of recycled materials and any efforts to reduce packaging in the handling operation.

Generally, an **organic livestock production plan** includes:

1. a description of the sources of livestock;
2. a description of the production method; and
3. a description of livestock management plans for diet, disease, pests, breeding and related problems production issues, in compliance with these standards.
4. requirements of the crop production plan, as appropriate

Certifying bodies should ensure that their application forms include a description of the operation's record keeping system. The standards mention documents – the intent is that the types of documents used by the operator should be identified in the plan. These records should be available for review by organic inspectors at the annual inspection, but may not need to be attached to the plan.

Records must be kept for at least five years, consistent with international requirements for organic certification.

"Record Keeping and Identification" – This paragraph (4.4) in the Standards describes the records that must be kept by an operation. 'Input' generally refers to farm production, such as fertilizers or livestock health products, but can also mean ingredients and cleaning products – anything that is brought onsite in order to produce an organic product. This paragraph also introduces the concept of traceability, which is the basis of organic certification. The guarantee of organic certification is that a certain product was produced according to the standards, and that the product can be traced from consumer sale back to the farm where it was grown. These standards require an operator to confirm the audit trail (written records) of a product while it is in their ownership. 'Release of the product' refers to the point at which an operator no longer owns a product, whether it is in their possession or not.

Specific records are necessary to ensure traceability (also called the audit trail): receipts, invoices, bills of lading, sales journals and financial (tax) records. In particular, sales records must ensure the traceability of the product, including such information as batch or lot numbers, production dates, label SKU, livestock tags and slaughterhouse documentation, or any other invention that will allow the purchaser (and further consumers in the supply chain) of a product to identify the origin of the product. These records facilitate an 'input/output' verification; that is, to ensure that an operation has not produced more 'organic' product than is possible from the organic ingredients it purchased.

Organic and non-organic crops must be distinguishable from each other, regardless of their intended use (e.g. a non-organic cow for home use must be distinguishable from the herd of organic cows and steers). This is particularly important for livestock operations, where an organic animal can become non-organic the instant it is treated with a prohibited medication (Note: there are specific exceptions for dairy animals and for the use of parasiticides). A treated animal which is not removed from the herd and continues to be managed organically can never be sold as an organic meat animal.

Certifiers will want to ensure that every livestock operation has a plan to deal with non-organic animals that remain part of the herd. Livestock ear tags (leg rings and bands for poultry) are one method, but the operator must be prepared to regard any animal without a tag (tags can be lost) as non-organic. If an operator does not use an identification system for their herd, they must develop a method of positively identifying non-organic animals (or they should keep them separate forever). If treated animals are shipped immediately, the operator must have records to confirm this. Unless the herd is very small, the statement that "there are no non-organic animals" should be backed up with evidence that the operator has an excellent health program (using approved health products) and/or that treated animals are quickly shipped into the non-organic market.

Organic and non-organic crops that are indistinguishable cannot be grown by the same operation (see 5.1.2). This statement refers to the appearance of a product – it is not adequate to put non-organic apples in special bins and say that they are now distinguishable from organic apples of the same variety. If an operator grows both organic and non-organic crops, the operator must ensure that they are easily distinguishable – e.g. organic wheat and non-organic barley. In this example, the operator could not grow any organic barley on their operation. Here are some examples of organic and non-organic varieties of the same crops which could be visually distinguishable:

- Nugget and mature potatoes – in this case, a producer could grow the same variety, as the organic and non-organic potatoes would be distinguishable
- Visually distinct varieties – red onions and yellow onions; red potatoes and white potatoes; green apples and red apples

Manufactured organic products must be distinguishable from non-organic products. The most common way to achieve this is through the use of product labels. Bulk organic products must be labelled in such a way that they cannot be confused with non-organic products.

Appendix

D Acronyms & Glossary

Acronyms

CB	Certification Body or Certifier
CFIA	Canadian Food Inspection Agency
CGSB	Canadian General Standards Board
COG	Canadian Organic Growers
COO	Canada Organic Office
COR	Canada Organic Regime
COS	Canadian Organic Standards
CVB	Conformity Verification Body
GE	Genetic Engineering
GM, GMO	Genetically Modified Organism (see Genetic Engineering)
MSDS	Material Safety Data Sheet
OPR	Organic Products Regulation
PMRA	Pest Management Regulatory Agency
PSL	Permitted Substances Lists

Glossary

Affidavit

A written statement of facts that can be used to hold the signer accountable if they make a false statement.

Canada Organic Office (COO)

The section of the Canadian Food Inspection Agency charged with enforcing the Canada Organic Regime

Canada Organic Regime (COR)

The overall process that includes definitions of:

- roles and responsibilities of various stakeholders;
- procedures for inspection and certification;
- conditions for use of the Canada Organic seal;
- surveillance and enforcement procedures;
- maintenance of the Canada Organic Standard and Permitted Substances Lists

Canadian Food Inspection Agency (CFIA)

Agency of the Canadian Federal Government that oversees and enforces all food related legislation www.inspection.gc.ca

Canadian General Standards Board (CGSB)

CGSB is a federal government organization that develops standards in support of the economic, regulatory, procurement, health, safety and environmental interests of government, industry and consumers. CGSB collaborates with organizations in various countries to develop national and international standards by being active in ISO (International Organization for Standardization) committees. www.tpsgc-pwgsc.gc.ca/cgsb/home/

Canadian Organic Growers (COG)
A membership-based national charitable organization whose mission is to lead local and national communities towards sustainable organic stewardship of land, food and fibre while respecting nature, upholding social justice and protecting natural resources. COG publishes handbooks for farmers, advises governments on policy and regulatory changes to support organic agriculture, and carries out many educational and advocacy projects. See page 3 and inside back cover. www.cog.ca

Canadian Organic Production Systems General Principles and Management Standards
"The" Standards for organic agriculture and food processing in Canada. Sections on definitions, organic plan, crops, livestock, processing, specialized operations (greenhouse, maple, honey, mushrooms, wildcrafting), emergency treatments, certification requirements, labelling requirements. Available from the Canadian General Standards Board at www.tpsgc-pwgsc.gc.ca/cgsb/ or by calling 800-665-CGSB

Canadian Organic Production Systems Permitted Substances Lists (PSL)
Part 2 of the Standards, the PSL lists the substances (for crops, livestock, processing, and cleaning) allowed in organic agriculture and food processing in Canada

Canadian Organic Standards (COS)
Also called "the Standards;" includes both parts, the General Principles and Management Standards and the Permitted Substances Lists

Certification Body (CB), Certifier
Organization that provides third-party inspection and certification services to organic growers and food processors, under the organic standards of various countries. A certifier can provide services for several national standards, for example certifying your operation to the Canadian Organic Standards (COS), the US National Organic Program (NOP), Bio-Suisse, etc; all from the same inspection. For a list of certifiers, see the CFIA website.

Committee on Organic Agriculture
Updates to the Standard are made by this committee (under CGSB). A group of technical experts from the organic industry that works through a consensus process to ensure the standards meet the needs of consumers, producers, general interest and regulatory groups.

Conformity Verification Body (CVB)
In agreement with the CFIA, CVBs assess, recommend and monitor the accreditation of certifiers. Formerly called Accreditation Body. For a list of CVBs, see the CFIA website.

General Principles and Management Standards
See Canadian Organic Production Systems General Principles and Management Standards

Genetic Engineering
A process of inserting new genetic information into existing cells in order to modify a specific organism for the purpose of changing one or more of its characteristics; in a way that does not occur under natural conditions

ISO (International Organization for Standardization)
The world's largest developer and publisher of international standards. This nongovernmental organization works with member countries around the world to establish consensus on standards, including national organic standards.

Material Safety Data Sheet (MSDS)
Internationally standardized way to document the hazardous properties of chemicals and other hazardous agents

On-farm processing
According to international organic standards, includes cleaning, extracting, packaging, blending, milling, dehydrating, cooking, juicing, and roasting, and applies to fruits, grains, herbs, honey, livestock products, maple syrup, vegetables, and other products at the farm. On-farm processing does not include washing and packing vegetables produced on the farm for sale at a farmers' market. If any off-farm raw products are used, the operation must complete "processor" paperwork and inspections.

Organic Plan (Organic System Plan, Organic Farm Plan)
A set of procedures for your operation. The Standards require you to prepare an annual Organic Plan, normally submitted as your application to a certifier for organic (re-)certification. Your ongoing records show how you implemented your procedures.

Organic Products Regulations (OPR)
An annex to the *Canada Agricultural Products Act* enacted on 11 June 2009 to make organic certification mandatory for trade across provincial or international boundaries.

Organic Trade Association (OTA)
A membership-based business association for the organic industry in North America. OTA's mission is to promote and protect organic trade to benefit the environment, farmers, the public, and the economy. www.ota.com/otacanada.html

Permitted Substances Lists
See Canadian Organic Production Systems Permitted Substances Lists

Pest Management Regulatory Agency
(PMRA), part of Health Canada, is responsible for pesticide regulation to ensure that pesticides pose minimal risk to human health and the environment.

Standards Council of Canada (SCC)
Because the organic standard is national, it has to be approved by SCC, which ensures that it complies with national and international trade rules. SCC is a federal Crown corporation that reports to Parliament through the Minister of Industry and oversees Canada's National Standards System. SCC approval gives the standards international recognition. www.scc.ca/en/index.shtml